Agricultural Tractors: A World Industry Study

Agricultural Tractors: A World Industry Study

Robert T. Kudrle

Ballinger Publishing Company ● Cambridge, Mass.
A Subsidiary of J.B. Lippincott Company

 This book is printed on recycled paper.

International Standard Book Number: 0–88410–034–0

Library of Congress Catalog Card Number: 75-20444

Printed in the United States of America

Library of Congress Cataloging in Publication Data

Kudrle, Robert T
    Agricultural tractors: a world industry study
    Originally presented as the author's thesis, Harvard.
    Bibliography: p.
    1. Tractor industry. I. Title.
HD9710.A2K82   1975   338.4'7'629225   75–20444
ISBN 0–88410–034–0

To My Parents

Contents

List of Figures

List of Tables

Acknowledgments

So many people have helped me during the preparation of this study that it is necessary to choose between an extremely long but inevitably incomplete list, and a very short one. The latter course seems both more practical and less likely to make the author appear to be an ingrate. While I was searching for a suitable dissertation topic in the Harvard Department of Economics, Raymond Vernon suggested that the studies of the Canadian Royal Commission on Farm Machinery could provide much of the material necessary for an unusually comprehensive industry study. Subsequently, Neil MacDonald, Director of Research of the RCFM, was helpful and encouraging in every way. Having found an exciting subject to investigate, it was my good fortune to persuade Richard Caves to direct my doctoral research. His endless patience and kindness are largely responsible for whatever is of value in the study. In the final stages of the project Marc Roberts provided extremely careful and invaluable criticism of the entire work. Many ribbons were shredded in the preparation of what follows; special thanks in this area are owed to Coral Sullivan, Cynthia Clapp, Carolyn McCann, and Penelope Roberts. Finally, no words can express my debt to Venetia H.M. Kudrle who helped with and shared in it all.

Chapter One

Introduction

The wheeled tractor is the most important piece of machinery in modern agriculture; an understanding of the structure, conduct, and performance [1] of the industry providing this vital machine is therefore of considerable importance. What follows, however, is not the usual industrial organization treatment of a single national industry but rather an investigation of the tractor industry in the seven major non-Communist tractor-using countries, ones which contained over four-fifths of all tractors in use in 1970,[2] over 60 percent of sales, and in which firms accounting for over 90 percent of all non-Communist tractor production were located (see Table 1-1). Because the tractor was not in widespread use outside of North America until after World War II, the study will be almost exclusively concerned with the post-war period.

THE FOCUS OF THE STUDY

The decision to focus on a rather narrowly defined product while at the same time investigating its production and sale over a wide, but not universal, geographic area both need preliminary justification and defense. The raison d'être of the study of industrial organization as it has developed over the past several decades is the evaluation of the production, distribution, and sale of a product, or product group, of sufficient homogeneity that questions about price-cost margins, the realization of scale economies in production, progressiveness in product design and manufacture, and quality and variety can be sensibly dealt with together. Our concern is with such considerations as they apply to the wheeled agricultural tractor, and perhaps least obvious is why it is thought appropriate to attempt to evaluate the performance of the industry in many different markets in a single study. Such a venture can be defended on at least two grounds: 1) through contrast and comparison, more might be understood about the industry in any one market than would otherwise be the case, and 2) inter-

Table 1-1.   The Seven Largest Wheeled Tractor Parks in the Non-Communist World—1970[a]

	1970 Tractor Park	1970 Domestic Sales
United States	4,600,000	176,637
Germany	1,369,662	61,798
France	1,240,000	64,250
Canada	549,789[b]	17,536[c]
Italy	482,341	35,500
United Kingdom	339,080[d]	32,802
Australia	303,399[d]	15,252
Total	8,884,271	403,775
Non-Communist World Total	10,535,082	653,800

[a]Park figures for agriculture only.
[b]1961—all tractors used in agriculture.
[c]Agriculture only.
[d]1969.
Source: Tractor park figures from *Food and Agricultural Organization Yearbook* (Rome: United Nations, 1971), pp. 484–490. Domestic sales from material cited for Figure 2–1. Total sales estimates from the Ford Motor Company.

action among the various markets might be such that only a multi-market study could really make clear the conditions bearing upon any one of them. The performance of tractor provision and after-sales service is certainly important enough to warrant investigation for any major modern agricultural country, but only if there are important gains from comparisons is a multi-market study justified, and only if there is significant international interaction (or possibly potential interaction) among providers are we really justified in speaking of a "world industry." The position to be developed here is that both conditions hold for the product under discussion.

ALTERNATIVE CONCEPTS OF INDUSTRY

There are at least three broad ways of thinking about the term "industry." The simplest takes as a starting point a product "ideally bounded on all sides by a marked gap between itself and its closest substitutes"[3] and defines the industry as "sellers who supply close substitute outputs to a common group of buyers."[4] This contrasts with two other definitions which concentrate either on the mutual dependence recognized among various firms [5] or on production similarities among firms, whether or not they actually produce the same "product" by the first definition.[6]

Wheeled Tractors as a Product

Is the wheeled agricultural tractor a "product" sharply enough distinguished from other products and sufficiently homogeneous internally to make it useful as an object of investigation? Evidence suggests that it is. Over a broad range of relative input prices, the wheeled tractor is the motive unit of choice for a wide variety of agricultural purposes. Originally, tractors were substituted for draught animals, usually horses. This process was virtually complete in North America before World War II; in the other countries in the study the process took place in the first post-war decade (except for Italy where the changeover was slower)[7] as versatile small tractors and government support for tractor purchase became widely available. Where landholdings have been extremely small, walking tractors have sometimes been employed, and in the hillier areas of Italy and marshy areas elsewhere, crawlers are found to be indispensable. Crawlers, however, are much too expensive to purchase, run, and maintain to be close substitutes in most agricultural situations.

The agricultural wheeled tractor with minor modification is singularly well suited to certain non-farm uses as well. The biggest non-farm market is North America, and the introduction of the modified agricultural tractor with shovel, backhoe, and other attachments in the early fifties was a means of escaping the cost and relative immobility of conventional construction equipment in many urban light construction settings. Another increasingly important non-farm use in recent years has been in mowing the grassy areas which surround major highway systems. In both agricultural and industrial applications wheeled tractors of different sizes and those produced by different makers (including those located in different countries) have been observed to be close substitutes for one another. It is therefore asserted, subject to further confirmation in Chapter Four, that wheeled agricultural tractors do possess the characteristics necessary for them to be usefully treated as a product in the sense of gaps in a chain of substitutes.

Industry as Oligopolistic Interdependence

Is there a particular set of firms not presently selling a certain product to the set of buyers served by established sellers which the latter can identify and with which there is mutually regarding behavior? This is the question which the approach to industry stressing oligopolistic interdependence attempts to highlight.

Richard Caves has presented one possible reason for the study of some industrial activity worldwide based on oligopolistic interdependence and compares some international with regionally segmented national activity:

> ... the subject of industrial organization ... deals predominantly with oligopolistic markets in which the sellers take some account of the influence of their actions on each other and on the outcome of the market as a whole. The reach of these perceptions about the seller's influence on his market thus is one of the principal criteria guiding our judgment about the extent to which the national market is fragmented into submarkets .... In a national market sellers respond to one another's actions in ways that

normally affect the whole market. Where the market is segmented, these interactions will occur sometimes in the national market, sometimes only in regional submarkets . . . . In the purely regional or local industry the interdependence among sellers is perceived only in the local market.

Is there some point in blowing up this model of the regionally segmented industry to the world level, deleting the word "region" and substituting "nation"? . . . For the concept to have much explanatory value the firms involved must to some extent recognize their specific interdependence across national boundaries . . . .[8]

It will be argued in this study that this kind of mutual dependence recognized across national boundaries does exist for the major tractor makers and has existed over most of the post-war period. One strong piece of *prima facie* evidence that this kind of mutual dependence recognized is important in the tractor industry is the actual extent of seller interpenetration among the major markets. A look at the seven major tractor-using countries reveals that by the mid-sixties, the three largest producers of wheeled tractors—Massey-Ferguson, International Harvester, and Ford—had non-negligible shares in all seven countries, while the fourth and fifth largest producers, Deere and Fiat, were active participants in five and four countries, respectively. In 1966, although the strength of the firms varied greatly over the seven countries, their combined share in every one was impressive. It totalled only 32 percent in Germany, but was 60 percent in France, 62 percent in Italy, 72 percent in the United States and Canada, over 75 percent in Australia, and 80 percent in the United Kingdom (see Tables 1-2, and 1-3).

Estimated economies of scale will be seen to be an important force behind multi-market sales by the major firms. The pioneering questionnaire study employed by Bain in the early fifties determined that one plant of optimum scale in tractor production might require an annual output of 90,000 units, or between a third and a half of United States requirements in the sixties. Although Bain's material suggested a rather flat long-run average cost curve, recently completed work by the Canadian Royal Commission on Farm Machinery (RCFM) found that costs drop over 18 percent between 20,000 and 90,000 units and that even then all scale economies are not exhausted using best practice techniques in either the U.S. or the United Kingdom. Ninety thousand units exceeds the domestic absorption of every country in the study except the U.S., and the total non-Communist world could be served by no more than eight plants of such scale.[9]

Table 1-4 reproduces estimates of production by firm and country in 1966. It should be noted that, although Massey-Ferguson, the largest producer, is a Canadian-based firm, all of its North American tractor production comes from plants in the United States, as does much of the rest of its output.

The leading firms in the tractor industry are among the largest in the world.

Table 1-2. Concentration in the Seven Selected Countries—1966

The U.S. and Canada			The U.K.		
	%	Units Sold		% (1964)	Units Sold
IH	23.0	59931	M–F	38.2	16539
Deere	22.2	57846	Ford	33.3	14417
M–F	14.4	37522	IH	8.0	3646
Ford	14.3	37261	BLMC (Nuffield)	8.0	3464
Four firm total	73.9	192560	Four firm total	87.5	37884
Case	6.5	16937	David Brown	5.0	2156
Allis-Chalmers	5.7	14852			
Oliver (Cockshutt)	5.6	14592			
Minneapolis-Moline	2.6	6775			
Eight firm total	94.3	245716			

France			Germany		
	%	Units Sold		%	Units Sold
M–F	21.8	17663	Deutz	19.6	14840
Renault	17.2	13936	Fendt	13.0	9843
Fiat	12.7	10290	IH	12.5	9464
IH	11.8	9561	M–F	7.8	5906
Four firm total	63.5	51450	Four firm total	52.9	40053
Ford	9.6	7778	Eicher	7.0	5300
Deutz	7.2	5834	Hanomag	5.8	3786
Deere-Lanz	4.0	3241	Ford	5.0	3559
David Brown	3.0	2431	Deere-Lanz	4.7	1590
Eight firm total	91.3	73975	Eight firm total	75.4	54288
			Fiat	1.7	1287

Italy			Australia		
	% (1964)	Units Sold		%	Units Sold
Fiat-OM	40.1	15915	M–F	23.3	4339
Same	15.9	6311	IH	23.3	4339
M–F (Landini)	13.4	5318	Ford	23.3	4339
Ford	7.4	2937	Three firm total	70.0	13017
Four firm total	76.8	30481	Fiat	5.0	379
IH	1.5	595			

Source: The market shares for the U.S., U.K., and Australia are estimates only. The U.S. and Canadian figures rely upon the output estimates presented in Table 1–4 and are based on the assumption that the foreign/domestic sales of all firms are in the same proportion as North American exports to North American production. The U.K. figures are from Donaldson, Lufkin and Jenrette, Inc., *The European Agricultural Equipment Industry and the Competitive Positions of North American Producers* (New York: Donaldson, Lufkin, and Jenrette, Inc. 1966) (Mimeographed.), p. 10. Material for the other markets is cited in the footnotes to Figures 7–3, 7–4, and 7–5.

Table 1-3. Total Firm Sales in the Study Markets—1966 (Unit Sales)

	MF	*IH*	*Ford*	*Deere*	*Fiat*
U.S.–Canada	37522	59931	37261	57846	– [a]
U.K.	16539	3464	14417	–	–
France	17663	9561	7778	3241	10290
Germany	5906	9464	3786	3559	1287
Italy	5318	595	2937	–	15915
Australia	4339	4339	4339	–	379
	87287	87354	70518	64646	27871

For all of the markets together:

IH	16.8%
M–F	16.8%
Ford	13.6%
Deere	12.4%
Fiat	5.4%

[a]In North America some models were sold by White.
Source: Calculated from material in Table 1-2.

In 1967 Ford was third in total world sales, International Harvester 30th, Fiat 46th, Renault 61st, Deere 107th, and Massey-Ferguson 169th. British Motors (80th) and Leyland (158th) merged in 1968. Massey-Ferguson and Deere are principally farm equipment manufacturers (75 percent of sales or more) which have recently moved into light construction, lawn, and recreation equipment as well. International Harvester does about one-third of its business in farm equipment and the rest mainly in trucks. Ford, Fiat, Renault, and British Leyland are principally automobile and truck firms, with tractors comprising only a small fraction of their total sales. Ford's enormous size is suggested by the fact that although it was the second largest wheeled tractor manufacturer in the world in 1966, tractors accounted for only about 3 percent of its global sales.[10]

As Tables 1-2, 1-3, and 1-4 reveal, not only is there a considerable gap in output between the top four firms and Fiat (after which the next largest producers, Renault and David Brown, are only half Fiat's size) but there is also a great difference among the top five in the extent of participation in different markets. Fifth-place Fiat and fourth-place Deere were noticeably more home market-oriented than the larger sellers in 1966 in terms of the distribution of total firm output. Our discussion of the world industry will necessarily treat the activities of these firms far more extensively than any of the smaller ones, although our investigation of the individual markets will deal with the activities of the smaller firms as well.

This study is confined to the seven major tractor-using countries in the non-Communist world. Diminishing returns from the addition of smaller users dictate such a cut-off point for comparative and seller-interactive investigation, but the study which follows is not comprehensive for yet another important reason. The very large volume of farm tractors produced and used in the Soviet Union and

Table 1–4. World Production of Wheeled Tractors, Actual and Estimated, 1966 (Except U.S.S.R., China, and East European Countries) (Thousands of units; underlined figures are estimates)

Company Ranking According to Market Share	World	U.S.A.	Britain	Federal Republic of Germany	France	Italy	Belgium	Sweden	Spain	India	Austria	Australia	Japan	Brazil	Finland	Others
Massey-Ferguson	153.8	38.8	78.6		29.2	3.2								4.0		
Ford	118.4	38.6	57.1				22.7									
International Harvester	108.0	62.0	21.0	15.0	8.5							1.5				
Deere	78.0	60.0		18.0												
Fiat (Fiat + Someca)	41.5				6.5	35.0										
Renault/Porsche	19.0				19.0											
David Brown	18.5		18.5													
J.I. Case	17.5	17.5														
Deutz	17.0			17.0												
Allis-Chalmers	15.5	15.5														
Brit. Leyland (Nuffield)	15.0		15.0													
Volvo	14.7							14.7								
Oliver (Cockshutt)	15.0	15.0														
Minneapolis-Moline	7.0	7.0														
Valmet	4.0													.9	3.1	
Other (known companies)	9.4								6.0	3.4						
Other (not identified)	157.2	15.6	20.2	51.0	2.1	10.8			7.1	8.6	11.7	9.3	9.7	1.1		10.0
World Total	809.5	270.0	210.4	101.0	65.3	49.0	22.7	14.7	13.1	12.0	11.7	10.8	9.7	6.0	3.1	10.0

Source: Clarence Barber (Royal Commission on Farm Machinery), *Special Report on Prices* (Ottawa: Queen's Printer, 1969), pp. 4–5.

Eastern Europe [11] (comparable in magnitude to that of the West in the late sixties) will receive only peripheral treatment. Too little is known of the physical characteristics of the greater part of current output in these countries to be able to evaluate the product properly, and the whole scheme of industrial organization analysis is designed to evaluate economic performance in market economies. Our treatment of Eastern activity must therefore be limited to its impact on the markets in which we can adequately focus our performance criteria.

Industry Approached Through Substitution
in Production

Our example of oligopolistic interdependence among sets of firms selling the same product to differing sets of buyers is a possible use of the term industry, but the logic of the argument could, of course, also apply between any small group of established sellers and any conspicuously likely small group of potential competitors. The prospects of the latter might well be based on the ease of moving into production of the product under consideration by firms serving approximately the same set of buyers with other products, rather than the ease of the geographical extension of sales (and perhaps production) of the same product. Such ease of substitution on the seller side of the market, whether involving oligopoly or not, lies behind much of the industrial classification in published government satistics and provides a third approach in defining an industry.

Substitution on the supply side is quite important for an examination of wheeled tractors for two reasons. First, while the substitutability in use among wheeled tractors is quite high from one power class to another as one moves from, for example, a 30 horsepower (h.p.) tractor to one of 50 h.p., the elasticity of substitution between a tractor of 30 h.p. and one of 120 h.p. is undoubtedly quite low. Furthermore, as will be seen in the following chapters, some sellers have produced tractor power sizes that do not substantially overlap. Our treatment of all of these firms as being in the same industry is in fact largely based on the similarity of manufacturing techniques which will be given close examination in Chapter Three.

Another important reason for considering the industry carefully from the standpoint of substitution on the supply side relates to the oversimplification which the term "supply" embodies. Supply can be far more than merely production, and it certainly is in the tractor industry. What a tractor user is concerned with is tractor *services,* as will be established below, particularly in Chapter Five. What this implies is that established channels of both distribution and after-sales service are of utmost importance in the industry. This is particularly crucial where sales are dispersed thinly over a wide geographic area. A supplier of various kinds of farm machines other than tractors, some of them perhaps power-driven, which had an established sales and service network appropriately located, might be a more formidable potential competitor to established sellers than a firm already producing more similar physical products. Furthermore, as will be established in Chapter Three, there are economies of scale in distributing a wide range of farm machinery under such circumstances.

This raises the question of whether or not our focus of attention might not more sensibly be the farm machinery industry than the tractor industry. Such an approach, which might be quite compelling if only North America were our geographic area of concern, will not be taken for a number of reasons. First, although most tractors are typically sold through the same retail outlet as at least some other farm equipment in all countries, only in the sixties did the offering of a comprehensive line of equipment to a common set of dealers by a single manufacturer become a critically important practice in most markets outside of North America. In Europe until recently tractors were often the products of the automobile industry, while implements (including self-propelled combines) were produced and sold by other firms.

Second, the argument for inclusive treatment of farm equipment on the basis of production similarities is not a strong one. Only the combine rivals the tractor in size and complexity and, except for the engine, it has an almost totally different design.

Third, an attempt to do a worldwide study of tractors plus all other equipment sold through the same final retail outlet would in fact entail an evaluation of the structure, conduct, and performance of the producers of a considerably different mix of equipment in different countries—tractors are by far the most universal major farm machine. In the late sixties in the United States, tractors constituted 35 percent of all farm machinery and equipment expenditure, in Canada 30 percent, in the U.K. about 40 percent, on the Continent approximately 50 percent, and in Australia 37 percent.[12] In many markets the combine is the second most important machine and, when self-propelled, its unit cost is usually greater than that of a tractor. The importance of the combine varies greatly by country, however, and among the lesser pieces of equipment, international variation is even greater.

There is a final difficulty. Not only are tractors unique among farm machines in having been evaluated for economies of scale in manufacture, but they are also the only major machine widely tested against internationally accepted performance criteria.[13] An international industry study must be careful to make certain that the goods sold in different countries can be meaningfully compared. Farm equipment is therefore rejected as an object of study on both theoretical and practical grounds, although tractors considered as the keystones of "full-lines" of farm equipment will sometimes necessarily be the focus of our attention, especially when discussing North America.

Conclusion on Industry Concepts

All three concepts of industry have something to offer in directing our attention to important considerations. Nevertheless, given the primary goal of industrial organization as the evaluation of industrial performance, it seems useful to organize an investigation of geographically diverse seller-buyer inter-actions so that variations in performance can be most clearly identified and analyzed. Where a common group of buyers and sellers interact to produce a distinct outcome in terms of price, product, and conditions of sale, we shall use the term market, and the term industry will be used to discuss purveyors to such a market. Nonetheless, it will be discovered that the conduct and performance of industries in markets so defined are heavily interdependent, both because of common production for many markets by several firms and because of mutual dependence recognized across market boundaries. We will therefore deal very extensively with the interdependencies of the world industry.

SPECIAL PROBLEMS AND OPPORTUNITIES OF WORLD INDUSTRY STUDY

Selecting the Unit of Analysis

If the treatment of buyers as a common group is accepted as a good starting point for investigation, the important question immediately becomes whether or not the national boundaries of the seven countries examined in this study pro-vide appropriate lines of demarcation, especially because throughout the postwar period both the United States and Canada have had no official commercial barriers to the entry of foreign-made tractors for farm use. The evidence suggests that, while tractor-makers have generally treated the United States and Canada as virtually one market, the variation of price and product among the six re-maining entities, due to differences in demand and both government and company trade barriers, render their separate consideration appropriate. The European Economic Community (EEC) may ultimately render the separate con-sideration of member countries obsolete, but this was certainly not true as late as 1970.

There are a number of additional issues which demand special attention in re-lating the categories of industrial organization to distinct yet allied national markets; the following seem especially in need of preliminary discussion.

Concentration

If it is accepted that in the tractor industry the nation-state usually con-stitutes the appropriate market, concentration in each such market is of interest as a factor contributing to the ability of existing sellers to exploit barriers to entry against non-participant foreign firms as well as potential domestic competi-tors. Overall sales concentration in the world industry is also important because it makes more possible actions based on the recognition of international mutual dependence such as firms avoiding competition in each other's established markets or the division of new markets.[14]

It is tempting to begin this study with a detailed discussion of concentration by market and overall. The story which follows is a complex one, and, although something resembling a structural summary might be helpful, it is the writer's position that to present such developments without explanation of the behavior (particularly entry and subsequent strategy) with which they are intertwined would be a rather pointless (and confusing) exercise. On the other hand, to explain in detail the determinants of changes in concentration would be virtually to summarize the rest of the work, and the story is simply too complex to be briefly outlined. The compromise chosen is reflected in Table 1–5 which presents concentration material in summary form for all of the markets in the study and in Figure 1-1 which shows the role in total non-Communist sales of the "big four" over the post-war period and which very closely parallels the aggregate fortunes of the firms in the seven countries taken together. Detailed market shares by firm will be presented in Chapter Seven.

In terms of Bain's classificatory scheme,[15] the British market has been "very highly" concentrated and the North American and Australian markets "highly" concentrated throughout, while the Italian market moved from rather low concentration to "high" by 1960. The French market attained "high-moderate"

Table 1-5. Concentration in the Six Major Markets—Selected Years

		1950	*1955*	*1960*	*1965*	*1970*
North America	Largest				(1966)	
	Firm	30	29	25	22	–
	4-Firm	75+	70+	69	72	–
	8-Firm	90+	90+	91	92	–
United Kingdom	Largest					
	Firm	50	35	40	37	35
	4-Firm	80+	80+	80+	85+	85+
	8-Firm	90+	90+	90+	90+	95+
France	Largest	(1951)				
	Firm	20	40	25	22	21
	4-Firm	52	68	75	64	61
	8-Firm	67	84	83	86	84
Germany	Largest		(1956)			
	Firm	–	13	16	21	19
	4-Firm	–	45	49	53	60
	8-Firm	50+	69	74	75	83
Italy	Largest		(1956)		(1964)	(1969)
	Firm	19	16	36	40	50+
	4-Firm	52	49	74	77	–
	8-Firm	65	73	84	87	–
Australia	Largest					(1969)
	Firm	27	34	–	–	23
	4-Firm	75+	75+	–	–	75+
	8-Firm	90+	90+	–	–	90+

Source: Material cited in Chapter Seven.

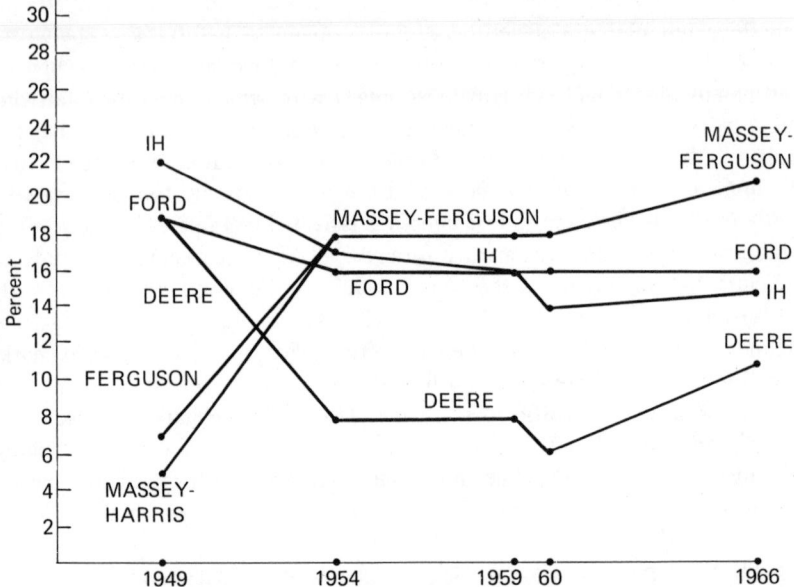

Figure 1-1. Estimated share of Non-Communist Tractor Sales, Selected Firms and Years

Note: Massey-Harris and Harry Ferguson merged in 1953. Data points were chosen on the basis of reliable estimates. Both 1959 and 1960 are shown to indicate the importance to International Harvester and Deere of the North American market and their vulnerability to low farm incomes there.

Source: Output for Ford is available from published company figures. See Mira Wilkins and Frank Ernest Hill, *American Business Abroad: Ford On Six Continents* (Detroit: Wayne State University Press, 1964), p. 439, and the Company's *Annual Reports.* Massey-Ferguson (and predecessor companies) has its tractor output presented in E.P. Neufeld, *A Global Corporation* (Toronto: The University of Toronto Press), pp. 61, 66, 121, 139, 284-285. Production estimates for all firms in 1966 are presented in Table 1-4. A U.S. production estimate for International Harvester in 1949 is from Michael Conant, "Competition In the Farm Machinery Industry," *Journal of Business,* 26 (January, 1953), p. 36. Other estimates are based on industry and official sources of production, market share, and trade cited in Chapters Seven and Eleven below.

concentration in the fifties, while the German market reached this level of concentration only by the mid-sixties.

Yet another measure, overall production concentration, as presented in Table 1-4, is important for understanding which firms may best be realizing scale economies, although it must be bolstered by detailed information on the degree of fabrication in each location. Finally, because worldwide fabrication may very well entail substantially differing unit cost at various locations, production location takes on importance in relation to seller shares in each market. A lack of correspondence between the cost of goods sold in the same market and particularly differential changes in those costs might have destabilizing effects on an oligopoly for which seller concentration alone would suggest substantial cohesion, while such differential costs or cost changes, if experienced on the fringe of the market, could have an impact that is scarcely noticeable, at least in the short run.

Conditions of Entry

Before attempting to discover what is especially important about the explicit consideration of "horizontal" international entry, it is necessary to confront the issue of how entry should be defined. The simple importation of a new brand of goods into a market by scattered interested buyers clearly is not entry in the active, calculated sense in which the term has meaning in industrial organization. On the other hand, identifying entry with "entry into production" seems unsatisfactory as a general criterion, although in many industries in many countries, some degree of fabrication in the market where sales are to take place is necessary for market penetration. Therefore, while the term "entry" in its usual usage implies both production and sales, the substantiality of a new firm's activity in sales competition, however accomplished, would seem the appropriate central object of attention. How active and direct a new firm's activities must be for it to be considered an entrant may vary considerably from industry to industry, but in a durable producer good industry in which after-sales service is of utmost importance, an appropriate criterion would seem to be the attempt by a new firm to set up a distribution and service system, whether through owned distributorships, supervised independent distributors, or both.

The discussion in Chapter Six will show that in an international industry study, with entry as here defined, conditions of entry require a more complex treatment than they usually receive. The reason for this is simple: industrial organization theory as usually presented does not allow for entrants to be better placed than existing sellers in any of the three areas of concern—product differentiation, economies of scale, or absolute cost. In the usual analysis, the position of the best placed established seller is taken as a baseline against which newcomers are at best at parity. When potential entrants may be relatively advantaged as well as disadvantaged according to the various considerations, it is the *net* position of the potential entrant which is of importance, and disadvantages are not immediately barriers. For example, it is intuitively obvious and will be developed in detail later, that an advantage in the level of factor cost can create an "absolute cost advantage" which can overcome any economies of scale, product differentiation, or other absolute cost disadvantage if it is great enough.

The possible advantages and disadvantages for an established firm entering from abroad over a competitor beginning de novo have been suggested by Caves.[16] Although his discussion is particularly concerned with direct investment involving production, the arguments apply with even greater force when entry is defined less stringently. The economies of scale problem is clearly overcome to the extent of commonality of the product to be introduced with production elsewhere. The product differentiation advantages of established firms in any one market may be considerably diminished by the existence of foreign producers who have marketing knowledge for the product in question and may even have established some familiarity among previously unreachable customers through marketing spillover. As noted above, an absolute cost advantage of importance for many industries may simply be the lower level of production costs available to potential entrants established elsewhere due to differences in factor costs, particularly efficiency wages. Other absolute cost advantages may include familiarity with production techniques and access to capital, either for production and distribution or for distribution alone, on terms more favorable than those available to potential domestic entrants.[17]

Against the added advantages of the foreign entrant must be weighed special disadvantages. The unique international absolute cost disadvantage for which, unlike transportation costs, there is usually no regional analogue, is, of course, the tariff or other trade controls, and not only the level of the tariff at any given time but uncertainties about its possible manipulation at the behest of internal interests could either deter entry or redirect it towards production within the new market. The latter course is also not without its special hazards. While the importing entrant may rely on domestic expertise in distribution, possibly in the form of established independent wholesalers, to compensate for some of his ignorance of the market, the producing foreigner faces additional informational problems concerning law, culture, and politics. These problems would clearly be more severe than those facing a native entrant. As information is costly, the foreigner may well settle for less of it and therefore put himself into a riskier position than an otherwise similar domestic entrant. Additionally, there is the possibility of political discrimination against foreign investors and the risks of exchange rate changes.

Weighing the special advantages against the disadvantages for the foreign entrant, it may be expected that in many industries in many countries the least disadvantaged entrant will be the foreign firm. Moreover, in terms of Bain's analysis,[18] it may well be that, in ranking potentially entering firms from least towards greater disadvantage, there will be a marked discontinuity after the leading established sellers in other markets have been identified. Further, in each market, leading established sellers may make somewhat similar determinations about each other. This kind of phenomenon will give rise to an important kind of mutual dependence recognized in the sense that a small number of firms worldwide may well perceive some symmetry in their positions vis-à-vis each other not based solely on participation in the same market or markets.

Conduct

The world industry study permits an exploration of the conduct of firms in different national industries and makes possible attempts to relate varying conduct to differing structure and performance on a comparative basis. In addition, such a study provides an opportunity to test two sets of hypotheses which have been developed in recent years. One set relates to the international corporation and the conduct patterns of world oligopolies and the other to suggested patterns of production, trade, and investment among industrial countries based on income levels, endowments, and tastes.[19]

Performance

The most obvious special opportunity afforded by world industry study for the examination of performance is the possibility it may offer for relating variations in performance to structural and behavioral differences among markets. Additionally, performance predictions can be related to some of the first set of hypotheses noted in the previous section. World industry analysis, however, in addition to providing a wider *scope* for the investigation of performance, might generate additional *criteria* as well. International differences in progressiveness in manufacture and in variety and quality of products might substitute for the otherwise missing test of good performance in these areas.[20] Another possible criterion is related to the conduct hypotheses concerning location of production and trade. If some world production locations have demonstrably lower unit costs at similar scales of output than others, should industrial organization take such cost differences into account in evaluating industry performance? The writer's tentative answer is "yes," while acknowledging that the demonstration of substantial cost differences over long periods of time would be the only way to apply such a criterion given the vagaries of differential international cost changes in domestic currency and changes in currency valuation.

EARLIER RESEARCH

Previous research on tractors has usually been as part of a larger investigation of the manufacture and sale of farm equipment. The Federal Trade Commission (FTC) published a 1,176-page study in 1938 and a brief follow-up a decade later to survey post-war conditions.[21] Much attention was paid in both studies to the allegedly anti-competitive effect of prevailing distribution channels. At approximately the same time as the second FTC study, the British group, Political and Economic Planning (PEP), published *Agricultural Machinery* [22] which not only surveyed the British domestic market but, largely in an attempt to evaluate export possibilities, dealt in some detail with demand and production conditions in other nations as well. Drawing heavily from, and expanding, the PEP study was the United Nations' *The European Tractor Industry in the Setting of the World Market* of 1952.[23] Both of these studies stressed the price and, by implication, cost advantage of British- over North American-produced machines. At about this time, three American doctoral dissertations were written about the North American farm equipment industry, though their attention focuses principally on the pre-war period. [24] The French manufacturers' association sponsored a comprehensive discussion of the French industry, *L'Industrie des Tracteurs et des Motoculteurs* in 1954.[25] In 1956, Phillips published his *Agricultural Implement Industry in Canada*, [26] an economic study which necessarily dealt rather extensively with the U.S., because a "common market" had been established for over a decade. In 1964, R.E. Linneman presented a marketing thesis which concentrated on the U.S.-based manufacturers' loss of overseas tractor markets.[27] In 1969, E.P. Neufeld published his thorough corporate biography of Massey-Ferguson, *A Global Corporation.* [28]

Until the 14-volume Canadian Royal Commission on Farm Machinery's studies and reports of 1969 through 1971, these were the only full-length studies addressed to farm equipment and tractors from a predominantly industrial point of view. There were also chapter-length studies by Whitney and Phillips on the U.S. farm equipment industry and one by Bernasek and Kubinski on the Australian industry.[29]

The work of the Royal Commission on Farm Machinery received considerable initial impetus from observed differences in the prices of nearly identical tractors in Canada and the U.K., but had many objectives. The several studies examined everything from safety and capacity to the attitude of farmers towards farm machinery purchases. The Commissioner (there was but one) was the well-known Canadian economist, Clarence Barber, and among the Commission's output was a great deal of material which was either written as economics or was at least grist for the mill of economic analysis.[30] One study, *Oligopoly in the Farm Machinery Industry* by David Schwartzman, is in fact an industrial organization study focused, however, almost exclusively on North America.

The author's debt to the Royal Commission's work and especially to the cooperation and encouragement of its Director of Research, Neil B. MacDonald, is immense. It is hoped that the present work can complement the Commission's efforts by focusing its attention exclusively on tractors, by treating factors outside of North America more extensively, and by putting the present situation of the world tractor industry into sharper historical focus. By doing so, our ability to evaluate performance and policy alternatives will hopefully be advanced.

PLAN OF THE STUDY

The behavioral developments which lie behind the sales and production patterns of Tables 1-1 through 1-4 together with some subsequent developments will be the concern of Chapters Seven through Eleven. They are preceded by discussion of wheeled tractor demand (Chapter Two); economies of scale, capital requirements, and international cost differences (Chapter Three); the development, differentiability, and differentiation of the product (Chapter Four); distribution systems (Chapter Five); and overall entry barriers (Chapter Six). The study concludes with an evaluation of the industry's performance (Chapter Twelve) and possible improvements which might be engendered by both private and public action (Chapter Thirteen).

The study is for the most part organized in the familiar way in which structure is examined first, followed by conduct and performance, but it should be stressed that unidirectional causation is not thereby implied. This is a point made by nearly all contemporary writers in industrial organization,[31] and it deserves emphasis here. For example, entry is at the same time the most fundamental act of conduct and, if successful, can result in substantial structural alteration. In some industrial organization studies, of course, this is an obvious but rather irrelevant consideration, because all principal actors are on the scene throughout, and entry-forestalling behavior is successfully pursued. Because the story told here is one of reaction to the threats and opportunities of larger and more numerous markets and sources of competition, however, the act of entry into a foreign market is a continuously crucial element of conduct. Further, as entry and the marketing strategies pursued thereafter have been such important determinants of concentration, the treatment below will deal with concentration, and especially changes in concentration, as a result of firm conduct as well as a determinant of it.

SOME PRELIMINARY QUESTIONS

In addition to the broad theoretical concerns already outlined, it might be appropriate to state in advance several specific questions about the tractor industry which the writer is particularly interested in answering. How have the major North American firms as a group maintained and even extended their world market share since the mid-fifties despite a large number of new international competitors which began substantial tractor operations after the war? Why, despite the enormous economies of scale estimated for the tractor industry, did approximately 65 percent of the free world's output in 1966 come from suboptimal plants after a post-war "shake-out" period of nearly two decades? Why, despite the relative diminution of North America as a production base (from 70 percent of all output in 1950 to 44 percent in 1955 and 28 percent in 1970), did such a large fraction of total output still originate from locations which had been proven to be as much as 25 percent or more higher in cost at any volume for at least 20 years than those elsewhere (Britain in particular)? How could price differences among different markets for identical or similar tractors, vastly exceeding those predicted on the basis of transport costs or trade barriers, be maintained year after year? Finally, despite the enormous advantages for established sellers (which remain to examined), why have the tractor-making operations of the largest firms not been more profitable as a whole than they appear to have been?

Chapter Two

Demand

Certain properties of demand are particularly important for industrial organization analysis.[1] The most important are usually considered to be price elasticity and the rate of growth of total demand. With similar structural characteristics otherwise, rapidly growing industries might be expected to have greater price competition than more stable ones within the same ranges of elasticity for the industry's product at any one time, because of the greater promise offered to a firm successfully expanding its market share; the firm is not necessarily made worse off in the ensuing period even if its price (or cost increasing product strategy) is matched by its rivals. Similarly, when comparing industries which differ only in price elasticity, the penalty paid for price-cutting and being followed is more severe where demand is inelastic, and the gains from cooperative pricing higher, so independent action should be less frequently pursued.

These conditions also have implications for the conditions of entry into an industry where scale economies are important relative to the size of the market, as is certainly the case in the tractor industry. A newcomer with no other relevant production must achieve a non-negligible share of the market in order to operate at minimum efficient scale (this assumes it is less than the entire market), and he must decide whether to enter at that scale, at one which is smaller, or not to enter at all. His expected fate upon entering will turn in large part upon the behavior of established firms, and there are two extreme assumptions he might make (assuming that only scale economies deter entry); one is that established firms will hold their output constant, allowing price to fall. Here high price elasticity is clearly favorable to entry in the sense of diminishing the extent to which established sellers can price above minimum long-run average cost. The problem with this assumption, which Bain dubs "pessimistic," in that established firms don't allow bygones be bygones and "move over," is that industry profits are not generally maximized. Thus a pricing strategy aimed at newcomers with "pessimistic" assumptions may not turn out to be credible as a deterrent. The other extreme assumption is that established firms hold their prices constant and reduce outputs to share with the new firm—the simplest assumption here is that output for the newcomer is passively determined at 1/n (n=number of market participants) of an unchanged industry output. In this case, there is no economy of scale barrier into the industry at all unless (or until) it is sufficiently crowded that the addition of one more firm will move all firms to a sub-optimal level of output. Furthermore, higher elasticity allows a higher maximum entry-deterring price against firms considering entry under these circumstances and, in this sense, contributes, ceteris paribus, to heightened rather than diminished barriers as under the previous assumption.[2]

The conjectural problem as discussed thus far is complicated enough, but it treats the very special case in which economies of scale provide the *only* barriers into the industry and which usually assumes that all going firms at any one time share output at least approximately equally unless some firms, notably entrants, purposely restrain output to avoid "disturbing" the market. This outcome results from the assumption that the product under consideration is at least roughly homogeneous. When other market conditions are introduced, particularly significant product differentiation as is the case in the tractor industry, the conjectural problem for a potential entrant becomes vastly more difficult.

On both extreme assumptions about the behavior of established firms, expanding industry demand is clearly favorable to entry. Where the output of the newcomer is expected to contribute to total industry output, the price fall is diminished by the rightwards shift of the industry demand curve. On the assumption that established firms hold fast to price, there is no way for them to charge above average cost and avoid continuous entry over time.

THE DETERMINANTS OF DEMAND FOR AGRICULTURAL TRACTORS

Of the two major sources of demand for wheeled tractors, agricultural and industrial, little investigation has been done about the determinants of demand in the latter category, despite its importance. In the late sixties it accounted for about 24 percent of U.S. sales, 10 percent in Canada, between 15 and 20 percent in Australia, and an unknown but significant fraction in Britain and on the Continent.[3] Nevertheless, although industrial demand in every market appears to be at least maintaining and perhaps increasing its unit share, our attention in this chapter will necessarily focus on agricultural demand.

The tractorization of agriculture outside of North America was almost wholly a post-war phenomenon. While there were already an estimated one and a half million tractors on North American farms in 1939, the tractor stock in each of the other countries was very small and similar to one another in total size. Britain had about 55,000 units, France 30,000, the part of Germany which was to become the Federal Republic, fewer than 50,000, Italy fewer than 40,000, and Australia 48,000.[4] Nevertheless, while French and German demand grew rapidly until the mid-fifties, the U.K., which was the only country outside of North America which substantially increased its park during the war,[5] saw only two post-war years of extremely high sales which were subsequently surpassed in the late fifties in response to government incentives; in Australia the 1950 sales figure was a high water mark. The U.S. and Canada experienced a post-war boom in home demand which resulted from replacement purchases long delayed, first by low farm incomes and then by war-time production controls. In both countries, however, 1953 sales figures were never again exceeded (see Figure 2-1).

Even when the measurement of tractor purchases is done on the basis of total value of sales rather than units, no market except Italy could be characterized as growing rapidly after the mid-fifties.

There have been a number of econometric studies attempting to establish the determinants of the purchase of tractors for farm use.[6] Tractors are a producer durable the usual economic life of which is between 10 and 20 years and the services of which are a derived demand. In a pioneering study, Griliches found that a very high percentage of the variability in the stock of U.S. tractors on farms over the years 1921–1957 could be explained by the relative price of tractors to crop prices received and the rate of interest at which rural loans were available. He did not establish a statistically significant relationship between desired tractor stock and the price of competing inputs. Subsequent studies, by Rayner and Cowling, and Scott and Smyth, however, using somewhat different estimation techniques and tractor stock measures, and using more recent data for the United Kingdom and Continental Europe, found the price of tractors relative to labor to be an important determinant of optimum tractor stock. Scott and Smyth found the elasticity of substitution between tractors and labor to be around 1.5. They also found farm size to be an important determinant of tractor demand. An increase of 10 percent in average farm size would, ceteris paribus, lead to a 5 to 6 percent rise in the ratio of tractors to labor (the smaller number is based on tractors measured in numbers, the latter measured in horse-power). The most likely explanation of this phenomenon was determined to be a divergence between relative prices of the factors and their relative marginal products. The assumption is that a farm family typically cannot hire out its own labor, and the farmer applies his own effort to a point where the marginal product is below the agricultural wage; thus the tractor-labor ratio in countries with unusually small farms (notably Germany) is less than implied by the relative tractor-labor price.[7] Despite the disequilibrium in rural labor markets, Scott and Smyth's results are entirely consistent with a rapidly diminishing farm labor force along with a virtually stagnant annual unit tractor demand, albeit for units of continuously greater horsepower.

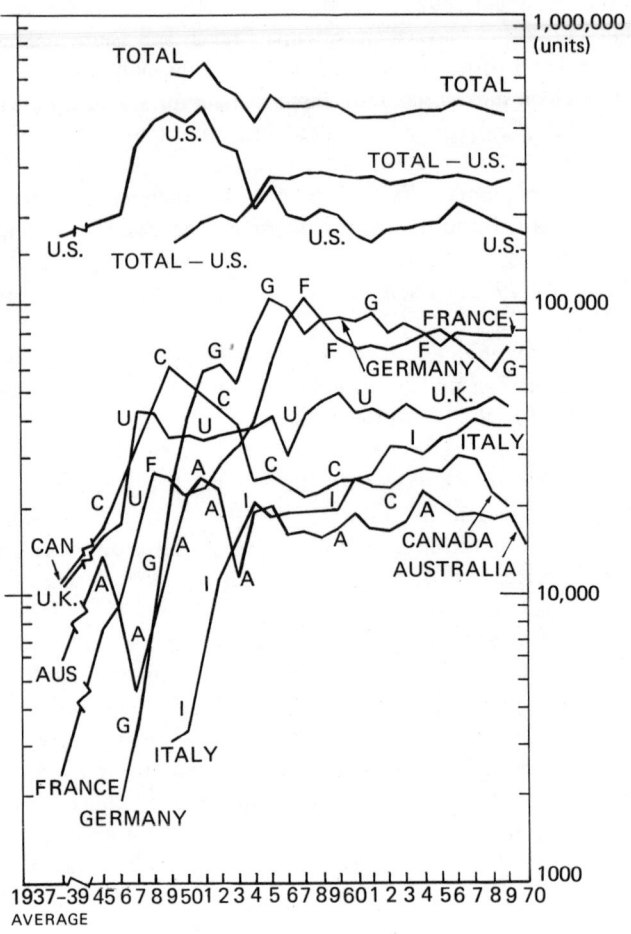

Figure 2-1. Tractor Demand, Selected Markets

Sources:

Total Demand:	Estimate of the Ford Motor Company cited in Barber, *Report*, p. 79.
U.S.:	Farm and Industrial Equipment Institute, *Farm and Industrial Equipment Facts* (Chicago: Farm and Industrial Equipment Institute), various annual issues.
Canada:	U.N., *European Tractor Industry;* Dominion Bureau of Statistics, *Farm Implement and Equipment Sales* (Ottawa: Queen's Printer), various annual issues.

Demand Elasticity

All of the econometric studies of tractor demand attempt to measure demand elasticity. Both stock and flow elasticities have been measured. An increase in desired tractor stock due to a decrease in the relative price of tractor services engenders an adjustment mechanism. Because the annual flow of new tractor services (usually measured either in units or horsepower) is small relative to the total stock (see Table 1-1), any given increase in induced demand measured over a discrete period will obviously be much greater relative to the previous flow than to the previous stock. It should also be clear, however, that the more gradual the adjustment mechanism, the less the previous flow will be disturbed. This implies that there is the greatest divergence between flow and long-run stock elasticities, ceteris paribus, when short- and long-run stock elasticities are close together. Both of Griliches' slightly different models produced a long-run stock price elasticity of -1.5 and a very low short-run elasticity (-.25) and annual proportional rate of adjustment towards desired stock (.17). The results of other studies vary considerably from those of Griliches in estimating greater responsiveness to price changes in the short run, but they all produce long-run elasticities between -1.0 and -1.8.

Although most of the estimates of long-run stock elasticity for the U.S. and Western Europe are closer to -1.5 than to -2.0, when looked at from a flow perspective the results of the studies are sometimes quite different. There one finds Rayner and Cowling's long-run stock elasticities of less than -1.5 corresponding to a short-run flow elasticity of gross investment (measured in physical units) of -2.3. Griliches' estimate is -1.9 for flow gross investment which is closer to his long-run stock elasticity of -1.5.

U.K.:	U.N., *European Tractor Industry;* A.J. Rayner, *An Econometric Analysis of the Demand for Farm Tractors;* the U.K. Ministry of Transport.
France:	Linneman, "U.S. Tractor Industry;" p. 40; Syndicat, *L'Industrie des Tracteurs;* C.N.E.E.M.A., *Bulletin d'Information,* various annual issues.
Germany:	*Landtechnik,* October, 1961, p. 658; U.N., *European Tractor Industry,* p. 34; *FIMR,* June 1, 1966, p. 657; *Landmaschinen Markt,* various issues (giving official registration statistics from the Statistiches Bundesamt since 1956).
Italy:	*La Meccanizzazione Agricola,* various years, and *AMJ* May, 1969, p. 49.
Australia:	Linneman, "U.S. Tractor Industry;" pp. 109–110; Commonwealth Bureau of Census and Statistics, *Rural Industries,* various annual issues.

Under some circumstances, the short-run elasticities just noted might tempt some firms to try to steal a march on their rivals, despite the presumably negative impact on the long-run profitability of the industry. Nevertheless, examining both growth of demand and elasticity together, one would not confidently predict active price competition, given the levels of concentration outlined in the previous chapter. Concentration was low on the Continent in the early post-war years, but European industry during this period was rife with agreements aimed at suppressing competition; it is therefore not surprising that price competition (discussed in Chapters Seven and Eight) has played only a limited role among the larger established participants in all of the countries in the study, and it also appears not to have been often used by small domestic firms, probably for reasons connected with their weak cost position due to small-scale production. Where price competition has been used is by foreign newcomers, although only in the case of those firms not representing a long-run competitive threat have the lower prices not been matched.

FLUCTUATIONS IN DEMAND

The short-run adjustment estimates presented in the previous section attempt to hold factors other than price changes constant, but, in fact, the most powerful short-run determinant of variation in tractor demand in the United States is the fluctuating income of the agricultural sector. There is no reason to believe that the U.S. is unusual in this regard, given the rather easily postponed nature of tractor purchase in all markets. The best estimates of the elasticity of annual tractor expenditures with respect to changes in net U.S. farm income lie in the range between .2 and over .5.[8] The important question here is how fluctuating demand would tend to influence the industry's conduct and performance. One important impact might be through widely varying plant utilization or the necessity of keeping very large inventories on hand. In fact, however, for the United States, sales in every year between 1956 and 1970 were within 20 percent of the average sales over that period (assuming no trend), and the following chapter estimates only very minor changes in unit cost accompanying changes in output within 20 percent of a plant's designed capacity. Inspection of Figure 2-1 reveals generally more minor fluctuations abroad, while Canada's pattern is similar to that of the U.S., and Australian fluctuations are somewhat greater. Because fluctuations within years as well as among them can contribute to production and inventory problems for an industry, it should also be stressed that by contrast with the sales of other farm implements (notably combines), tractor sales are not highly seasonal (their skewness in North America differs little from that of automobile purchases, although the seasonal pattern is somewhat different)[9] (see Table 2-1).

Table 2-1.   Seasonal Variability of Tractor Sales—Monthly Percentages

	U.S.	Canada	U.K.	France	Germany	Italy	Australia
January	7.0	2.4	U	9.5	8.9	4.0	
February	6.5	3.0	N	7.1	9.9	6.7	
March	9.6	5.3	A	8.1	11.5	10.6	25.4[b]
April	12.4	12.7	V	8.6	7.7	11.1	
May	12.5	16.3	A	8.8	7.7	12.4	
June	9.3	10.6	I	8.8	7.6	12.7	27.9[b]
July	6.4	6.3	L	7.1	7.1	10.0	
August	6.2	6.2	A	6.4	6.8	7.1	
September	7.6	7.9	B	6.7	9.0	9.1	21.4[b]
October	11.3	21.6[a]	L	8.5	8.6	9.3	
November	5.5	4.0	E	9.3	7.5	4.2	
December	5.5	3.6		11.0	8.1	2.7	25.3[b]

[a]Thought to include units sold in previous months.
[b]Three-month cumulative totals.
Source: U.S.: 1969, *Implement and Tractor*, November 21, 1970, p. 42.
Canada: 1965–67 average, Schwartzman, *Oligopoly*, Appendix A, Table A.5, p. 222.
France: November 1966–October 1967, Centre National d'Etudes et d'Experimentation de Machinisme Agricoles, *Etudes*, May, 1968, Tableau 1 [p. 1].
Germany: 1966, *Landmaschinen Markt*, May, 1967, p. 603.
Italy: 1958, Utenti Motori Agricole, *La Meccanizzazione Agricola in Italia–1958*, p. 207.
Australia: 1970, Commonwealth Bureau of Census and Statistics, *Sales of New Tractors, December Quarter*, 1970, p. 6.

INTERNATIONAL DIFFERENCES IN TRACTOR POWER AND FUEL USED

There has been an enormous growth in the size of the average tractor sold in North America in the post-war period and a less dramatic but important increase elsewhere. Ninety percent of the tractors sold in the United States in 1950 were of less than 35 h.p. and only 1.4 percent were above 50 h.p. In 1970 only 10 percent of sales in the U.S. were of units of less than 35 h.p. and 71 percent over 50 h.p. (and 36 percent over 70 h.p.). Canadian figures were similar. For other countries in the study, tractors were, if anything, smaller than those in North America at the war's end. Nevertheless by 1970 about 55 percent of all units sold in Britain were over 50 h.p. and only about 6 percent were under 30 h.p. In France in 1969, 45 percent of all tractors sold were 50 h.p. units or larger and 11 percent were under 34 h.p.; the comparable German figures are 35 percent and 33 percent for 1968. In 1968, Italian sales were 23 percent for units under 30 h.p. and 29.2 percent for units over 52 h.p. In Australia in 1970, only 4.0 percent of all sales were of units of less than 35 h.p. while 57 percent were of units of over 45 h.p.[10]

It is clear that average farm size explains much of the observed variation. Holdings in all countries have increased over the period, and while Italy has had the smallest average farms throughout, the average tractor purchased has been somewhat more powerful than that in Germany because of the hilly country in which many Italian tractors must work. French farms and tractors have been larger than those of Italy and Germany throughout, Britain's and Australia's larger yet, and those in North America the largest of all.[11]

While average farm size and the price of labor no doubt explain a good deal of international variation in the average horsepower of tractor purchased, the actual determinants of the average size of machine sold have never been subject to econometric investigation, as far as the writer is aware. Even less is known about the determinants of the structure of demand for machines of different sizes. This suggests an enormously complicated issue with implications for both new and used tractor markets. "Tractor service" is not only a derived demand but is itself an extremely heterogeneous concept from which the purchase of tractors, measured either in units or horsepower, is yet one step further removed. In North America, for example, increasing labor costs and the necessity for more rapid field operations to best take advantage of sophisticated ancillary inputs such as pesticides and fertilizers have led to a very large increase in the average horsepower of machines sold for farm use. This, however, is only part of the picture. A rapid pace in performing critical field operations is only one type of "service"; there are other routine chores around a farm for which the increasingly large units have become poorly suited because of their lack of maneuverability and high operating cost. Partly as a result of these considerations, there has been a secular increase in the number of tractors per farm, while the number of hours worked per tractor has diminished, the latter as a result not only of the use of different size tractors for different tasks but also because many functions which in earlier years were performed by tractors (such as propelling combines) have been taken over by more specialized machines.[12] By the end of the period under review these trends were probably importantly affecting some non-North American markets as well, notably those of Australia and the United Kingdom.

The international variation outlined above implies that although there was considerable overlap in the power of machines sold in all of the countries in every year, the bulk of German machines actually produced, for example, would have been unsuited to most of North American agriculture at any given time. Another difference has been in fuel preference. Although many tractors in all parts of the world used heavy cheap fuels before the war, in the post-war period the cost of these fuels in North America rose and nearly wiped out the advantage. All U.S. manufacturers opted in the first instance for regular gasoline. Gasoline costs relative to diesel were much more favorable than in most of the rest of the world—except for France—a fact which resulted in a rather small number of diesel units being produced and, consequently, their high cost. As the horsepower of an engine is increased, however, the lower running cost of a diesel unit by comparison with the lower initial cost of a gasoline unit begins to take on greater significance. Over the years in North America a combination of the demand for more powerful tractors, increasing economies of scale in the production of diesel engines, and improvements in diesel engine design making them easier to start contributed to a very great change in North American demand. By 1970, gasoline units were dominant only in the under-35 h.p. category in which only one tractor out of ten was sold. In that year 70 percent of all units sold were diesel-powered. [13]

Kerosene had been the most economical fuel in Australia before the war and powered virtually all of the machines in use, but during the post-war period both gasoline and diesel power grew in importance as relative fuel prices changed, and the diesel proportion grew much more rapidly; diesel sales were 80 percent by 1958 and 98 percent by 1969. [14]

In France an abrupt change in relative fuel prices engineered by the government in 1954 caused an almost total switch from gasoline to diesel fuel over the next five years. [15] In all of the other markets over the entire period, diesel units have accounted for almost all domestic sales. [16]

What is the significance for the industry of differing national fuel preference and different sized machines, especially between Europe and North America? One observer lays great stress on the fact that European machines in the fifties were unsuitable for most of U.S. demand because of their diesel engines and small size (and vice versa). [17] As an explanatory factor for the lack of trade between Europe and North America (see Chapter Eleven) this is rather weak. The technology was available and nearly static (although in the case of diesels improving somewhat) in all countries after the mid-fifties, there was no lack of easy availability of suitable engines of either type from many suppliers on both sides of the Atlantic, [18] and, as established in the following chapters, most large tractors are merely scaled up versions of smaller ones. Furthermore, because the vast majority of tractors sold in every market never varied more than slightly except for size, fuel, and hydraulics, a point to be substantiated in Chapter Four and examined in even greater detail in Chapter Ten, all of the markets could easily have been served from a single plant if only production considerations were relevant (the complicated distributional problems remain to be discussed in Chapters Five through Eight).

Not only would producing for the various markets not have been seriously hindered by discontinuities in international tractor design but it might well have tended to stabilize the total production output of the firms involved, hence contributing to the stability of profits, particularly where production was centralized. This appears to be confirmed by Figure 2-1.

PUBLIC POLICY

A final demand issue of significance in the conduct and performance of the industry is public policy toward agriculture and, in particular, governmental support for the purchase of tractors. A really complete survey of all of the various special financial incentives to agricultural mechanization would take us too far afield, however, and only ones important enough to have a powerful impact on tractor demand will be discussed.

The United States government has never taken any substantial specific steps solely to increase tractor purchase, but excessive optimism about the level of government agricultural supports contributed to vastly exaggerated industry expectations about unit tractor demand in the fifties. Industry sources have admitted in retrospect that only part of the misestimation resulted from an unexpectedly lower support for certain agricultural products while much of it resulted from the primitive (really nonexistent) industry procedures for projecting unit demand.[19] Figure 2-2 shows the projected annual production from 1955 to 1960 by comparison with that which actually took place.

In every other market in this study the government has directly supported the purchase of tractors at various times, as well as exerting a major influence on the receipts of the agricultural sector.

In Canada, a scheme was set up in 1944 which operated through the commercial banks and guaranteed the participating banks for losses up to 10 percent of loans made under the plan. As much as a quarter of all machinery sales may have been financed under the scheme in the immediate post-war years, but interest rate ceilings in subsequent years made the scheme far less attractive for the banks, and although interest ceilings were substantially adjusted upwards in 1969, this was not followed by a large increase in the loans which had virtually dried up in the sixties. The upward adjustment was apparently still not enough to make the loans attractive to lenders.[20]

In Britain, government policy encouraged tractor purchase by a very steep investment tax allowance system over most of the period which allowed as much as 55 percent of the cash price of a tractor to be deducted from the tax liability in the first year and 76 percent over three years. This presumably explains much of the unusually high ratio of annual sales to rather stagnant tractor park revealed in Table 1-1 and also the complaints of Continental dealers about the flood of used British machines. The scheme was abolished in 1969 and replaced by a capital grants system, the rationale of which was to aid farmers too poor to have the tax liability necessary to benefit from the previous program. Only 10 percent of British financing is done on an installment contract.[21]

On the Continent, credit for agricultural capital investment, including tractor purchase, has often been made available on terms much more favorable than those provided by the free market, although sometimes conditions for especially favorable treatment have included the consolidation of land holdings. Fuel subsidies and accelerated depreciation have also been employed. The most important sustained encouragement for tractor finance was the Italian "Fanfani Plan" which, beginning in 1952, allowed 75 percent of the price of an Italian-produced tractor to be borrowed at a 3 percent interest rate for five years. The plan continued into the sixties despite protest from Italy's EEC partners.[22]

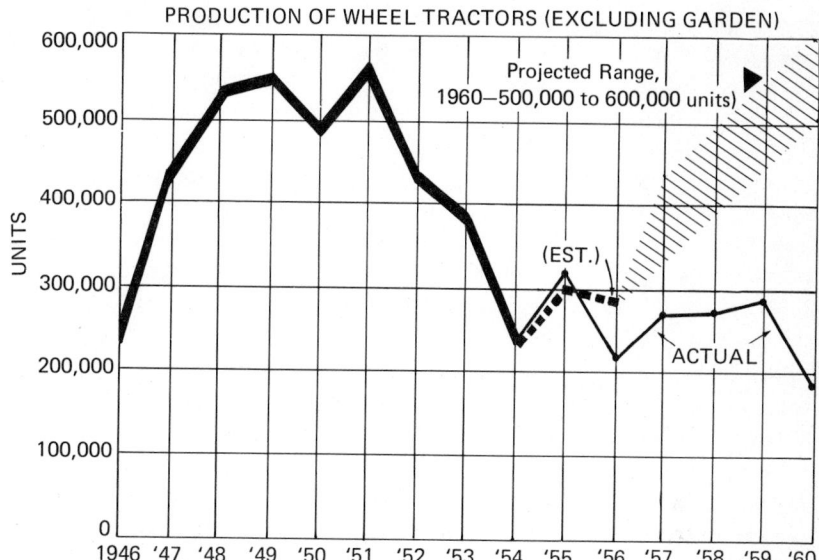

Figure 2-2. Implement and Tractor's Projections of U.S. Wheeled Tractor Production from 1955 Through 1960 and Actual Production
Source: Implement & Tractor, September 24, 1955, p. 116.

In Australia, in addition to tax concessions for agricultural capital expenditures, it appears that the Commonwealth Development Bank has been prepared to step in to assist in retail finance on very favorable terms, where other applications for loans "on reasonable and suitable terms" were not forthcoming.[23]

The role of the government in the retail financing of machinery would appear at first glance to be of considerable importance for this study, because most North American manufacturers have usually provided such finance in the home market. The importance of this provision as a necessary condition for participation in the North American market will be given a close look in Chapter Eight. Whatever its competitive importance in North America, however, it is clearly diminished elsewhere by government support of tractor purchase.

Chapter Three

Economies of Scale and Cost Estimates

A number of issues related to the cost of tractor production and distribution deserve careful attention. The first section of this chapter examines cost estimates and the economies of scale in North American tractor manufacturing in the late 1960s. This is followed by an attempt to estimate tractor costs in Britain during the same period. Scattered pieces of evidence about other countries' costs at various times are also presented. The chapter then discusses possible production complementarities with other goods and concludes with estimates of costs and economies of scale in distribution.

The estimation of economies of scale in the production and distribution of any good or service is important as an indication of how large a market a firm must have in order to operate at an efficient scale and how much efficiency is sacrificed by a lower level of production. If the estimates can be held with confidence, they allow the investigator to rate an industry on the basis of how much of its total production comes from plants (or firms, where there are economies of multi-plant operation) of efficient scale.

Of at least equal importance is the consideration of economies of scale as a barrier to entry. How much of the market must be wrested away from other competitors (on different assumptions about their price policy and hence total market size) in order for the entrant to operate at optimal scale, and how much does it suffer by operating at a smaller scale? Similar and interacting considerations apply to economies of distribution. Economies of production and distribution based on activities related to other products may be important, as are international activities with the same product. The latter consideration is discussed in Chapter Six.

ECONOMIES OF SCALE IN NORTH AMERICAN
TRACTOR MANUFACTURE

By far the most complete attempt to estimate both costs and economies of scale in the North American tractor industry is the Canadian Royal Commission on Farm Machinery's (RCFM) *Farm Tractor Production Costs: A Study in the Economies of Scale*, a formidable piece of work done during 1968 which developed complete paper plants for the production of three different sizes of farm tractors at three different output levels. The output mix (30 percent of plant volume at approximately 40 h.p., 60 percent at 90 h.p., and 10 percent at 130 h.p.) was based on demand conditions projected for North America at a time slightly later than those prevailing when the study was undertaken. One declared purpose of the study was to examine the prospects for new Canadian entrants into the North American industry; others were to use the cost estimates to evaluate the profitability of tractor manufacturing in North America at tractor prices prevailing in the late sixties and to compare the relative profitability of different models.

Three other Commission works lean heavily on the study's findings. *The Special Report on Prices* extends the analysis of the cost study to include larger volumes and British production costs, and David Schwartzman's *Oligopoly in the Farm Machinery Industry* and the Commissioner's (Clarence Barber's) *Report* both accept *Tractor Costs'* results and draw conclusions based on them. The study was a joint venture, with two members of the Commission staff joined by two specialists from Booz, Allen, and Hamilton Canada Limited, a consulting firm with extensive background in the planning of metal fabrication plants.

By comparison with an automobile, a tractor is a simple machine. While a car typically has around 15,000 parts,[1] a farm tractor has a mere 2,000.[2] Of these, 1,365 are seldom or never actually manufactured by the North American tractor-maker. Not only are virtually all non-metal (cast, machined, or stamped) parts purchased from outside suppliers, but in addition the tractor-maker finds it profitable to buy highly standardized metal parts (nuts, bolts, etc.) and even some large and complicated metal components such as radiators.

The Commission's procedure was to look carefully at the parts constituting about 70 percent of the in-house fabricated cost of a tractor when produced at a 60,000 per year volume. The best method of production and cost of these parts was found for three specific annual outputs: 20,000 ("small"), 60,000 ("medium"), and 90,000 ("large"). The costs of the remaining parts (often very similar to those looked at in detail) were then approximately estimated. Wage levels, shifts, hours, and levels of labor efficiency were keyed to actual industry conditions in 1968, and the latest proven technology was employed. While the unit costs estimated were those for a weighted-average size tractor, with the weights being those approximating North American demand in the late sixties, output levels were selected to represent the range of output most North American producers were actually experiencing at the time of the study.[3]

Tractor Costs assembles data on all the current and capital expenses necessary for the examination of the three volume-level tractor plants in very great detail. The longest lived piece of capital equipment is the building, which is assumed to have zero market value at the end of 20 years. Because economies of scale are usually considered to apply to the relation between average unit costs and output level per unit time, the problem becomes one of properly encapsulating capital costs in the analysis.

The costs shown in the first two columns of Table 3-1 are simply derived; the present value of all costs is divided by the present value of all tractor units produced over the life of the project, discounted at 10 percent and 8 percent, respectively. (The implied assumption that all variable and future capital outlays are to be made at "today's" level of cost is clearly inadequate, as is the assumption of a stable output mix, but the exercise is still instructive so long as its limitations are kept firmly in mind.) *Tractor Costs'* own calculations, shown in the third column of Table 3-1, are developed from a very different procedure. Instead of attempting any method of estimating representative tractor costs over the life of the project, the *Tractor Costs* team simply chooses one year—the third year of plant operations—for analysis. This year is justified by the suggestion that whatever initial "learning" problems might raise cost temporarily, the plant should be running at minimum cost by the third full year of operations. There is no further justification for the third as against any subsequent year. Capital costs are treated as the sum of straight-line depreciation and 7.5 percent of undepreciated capital.[4] This is a method which, of course, finds a different

Table 3-1. Production Costs and Economies of Scale (U.S. Dollars)

Annual Output	1 10%	2 8%	3 Commission
20,000 units	$3945	$3891	$3824
$ Difference	$ 440	$ 427	$ 412
% Cost drop	11.1%	10.9%	10.8%
60,000 units	$3508	$3465	$3472
$ Difference	$ 291	$ 289	$ 291
% Cost drop	8.2%	8.3%	8.5%
90,000 units	$3217	$3176	$3121
Total $ Difference	$ 731	$ 719	$ 708
% Cost drop	18.5%	18.5%	18.4%

Note: Columns 1 and 2 are the unit cost of an "average" tractor over the life of the project using discount rates of 10 and 8 percent, respectively. In Column 3 are the Commission's estimates explained in the text.

Source: Calculated from material in MacDonald, et al., *Tractor Costs.*

average cost for every year of plant operation and, with all market costs and prices constant, a much higher rate of profit at the end of the project than at the beginning.

Despite the difference in method, the discounting approach and the "point-in-time" approach yield very little difference in tractor unit costs and virtually none in economies of scale. Using the discounting method, an 8 percent rate produces unit costs about 1 percent lower than a 10 percent rate, and the 8 percent discounted costs are about 2 percent higher than those estimated by the Commission. Both methods establish that approximately 11 percent in cost reduction is achieved by expanding output from 20,000 to 60,000 units per year and a further 8 to 8.5 percent reduction is gained by moving to 90,000 units. Similar capital intensity and configuration of capital outlays over time for each size of plant account for the stability of estimated scale economies when the different estimation techniques are employed.

Using the somewhat arbitrary transfer price between manufacturing and distributing divisions within the industry actually prevailing in 1968 of $4,000 (U.S.) [5] (which corresponds to 58 percent of suggested retail price and 80 percent of wholesale price), the internal rate of return after taxes for tractor manufacture goes from 6 percent for the 20,000 unit operation to 16.5 percent at 60,000 and 22.5 percent at 90,000. The Commission discovers very similar rates of return on the basis of its own cost calculations. A discussion of these calculations in relation to actual profitability will be presented in Chapter Twelve.

Beyond the economies found between 20,000 and 90,000 units, both *Tractor Costs* and information gathered by the Commission from company experience for the *Special Report on Prices* imply that economies of scale persist for yet higher outputs. The industry sources seem to imply that increasing the yearly volume from 90,000 to 120,000 units per year would decrease cost on an average tractor by a further 5 percent.[6]

DISAGGREGATED ECONOMIES OF SCALE

The basic metal-working procedures used in tractor manufacture are very similar to those in the automobile industry: the casting of molten iron, the precision machining of those parts cast,[7] the stamping of sheet steel, and the final assembly of the in-house and outside-purchased parts.

Table 3-2 presents a breakdown of tractor unit costs estimated at an 8 percent discount rate for the three output levels studied by category of input. The changes which occur as volume expands require some explanation. The Commission found a dramatic rise in the degree of in-house fabrication between 20,000 and 90,000 units; as output increases, outside purchased parts expense drops by over 30 percent. This drop is in turn the result of two phenomena: the substitu-

Table 3-2. Total Unit Cost by Input and Plant Size (U.S. Dollars)

	Annual Unit Output					
	20,000	60,000	90,000	20,000	60,000 Index	90,000
Purchased parts	$2123	$1828	$1418	100	86	67
Basic material	362	397	482	100	110	133
Labor	398	385	416	100	92	105
Capital costs	465	394	397	100	85	85
Support costs (labor)	279	231	215	100	83	77
Operating expenses[a]	264	230	248	100	87	94
Total unit costs	$3891	$3465	$3176	100	89	82

[a]Including taxes and insurance.
Source: Estimates determined from material in *Tractor Costs* using an 8 percent discount rate over the 20-year life of the project.

tion of in-house production for outside purchased parts and a reduction in price of those parts which continued to be purchased. The increase in material expenses per tractor are also somewhat mitigated by economies attaching to large-scale purchase. Industry purchasing experts agree that overall purchased parts and materials would be about 9 percent cheaper at the 90,000 unit plant level than at the 20,000 unit level, holding the degree of vertical integration constant (at the degree appropriate to an annual output of 60,000 units). These industry sources ascribe most of the reduction not to monopsony power but rather to volume discounts relating to actual supplier savings and the ability of the larger and more specialized purchasing staffs of the larger plants to make more thorough and active ongoing searches of alternative sources of supply.[8]

The question still remains: why is it profitable to become more vertically integrated as volume rises? The Commission does not doubt its supplier price information,[9] so the answer in the case of much of the parts and materials, for which there is a large industrial demand elsewhere, must lie either with high supplier price-cost margins or higher supplier costs at even higher volumes than the new tractor facilities would attain, or some combination of the two.[10]

Direct labor costs go up by only 5 percent per unit over the range studied, although the in-house fabrication per tractor in a 90,000 unit plant is vastly greater than that in a 20,000 unit operation. Capital costs fall by 15 percent between 20,000 and 60,000 units and then remain stable. The modest increase in direct labor costs admits to no easy summary explanation, but the saving in capital costs is most dramatic in machining, where machine utilization is vastly increased at high outputs. Falling labor "support" costs (which are simply the sum of all wage and salary expenses not tied directly to production) are due largely to the insufficient divisibility of some inputs at low level of output.

It must be stressed that the estimates developed both in *Tractor Costs* and here are the costs involved in putting a finished tractor at the end of an assembly line at different annual rates; distribution, central administration, and the very considerable inventory costs (beyond those of goods in process) are not included.

THE COMMISSION'S ESTIMATES VERSUS BAIN'S ESTIMATES

Because this study is of the industry over time, it is important to compare the Commission's plant economies with those of Bain for the early fifties. His questionnaires revealed that the plant size judged optimal in 1951 or 1952 was perhaps "at the outside" 90,000 per year.[11] Where the Commission's findings diverge greatly from Bain's are in the shape of the long-run average cost curve. Bain's discussion implies that the 20,000 unit plant should experience costs not more than 5 percent higher than the 90,000 unit plant,[12] while the Commission found them to be 23 percent higher, and this cannot be due to any substantial extent to changes in technology; it has remained virtually unchanged over the period.[13] One can speculate that the boom conditions prevailing in the industry during the years for which the data were gathered and the prosperity of even the small firms in the industry might have caused some exaggeration of the flatness of the curve by respondents to Bain's questionnaire. It is also possible that the result was engendered by small firms operating more X-efficiently [14] at this time, although there is nothing in the record either at that time or since to suggest that this was the case.

It should also be noted that while Bain's "tractor" industry includes crawlers and contractors' off-highway tractors, there is no reason to believe that scale economies for these machines are different from those for wheeled tractors. It is not known from which firms the questionnaires were received or how responses were reconciled with one another.

Bain concluded that a plant of no more than 15 percent of national capacity would exhaust all significant economies of scale. Wheeled tractor output in the U.S. was 555,000 in 1949,[15] 500,000 in 1950, and 570,000 in 1951. These figures, however, reflected a great backlog of demand built up during the war. Wheeled output was less than 200,000 in 1939 and nearly down to that level again by 1956, the year *Barriers To New Competition* was published. After 1955, wheeled tractor output never exceeded 250,000 in one year and in several years dropped below 200,000, so that the optimum plant determined by Bain and the Commission's "large" plant corresponds more closely to between 35 and 50 percent than 15 percent of national output, and a yet higher percentage of national absorption.

Because an investigation of British tractor manufacturing and input costs revealed economies of scale in 1968 similar to those in North America, and, in addition, the newest British plant (Ford at Basildon) employed techniques which imply yet further scale economies (only briefly dealt with and unestimated for

North America in *Tractor Costs*), it is probable that economies of scale are extremely important in every Western country. It is therefore interesting to compare a 90,000 unit plant with home sales and production for each market. The difference between home sales and production itself, of course, represents a net trade balance, the determinants of which remain to be explained. The comparison is presented in Table 3-3. British absorption has been less than one medium plant's output throughout, although production in recent years has been about that of two large plants. The smallest 20,000 unit plant would have taken care of all of Australia's requirements in nearly all except the earliest post-war years, while actual production has been only a small fraction of that number. The output of one 60,000 unit plant would have greatly exceeded Canadian requirements over the entire period, although little production has ever taken place there. French production has been at the large plant level or lower, with absorption of similar magnitude. German production, too, has been at about the level of a large plant or somewhat higher, while absorption during the sixties was considerably lower. Although Italian production grew very rapidly in the sixties, it was still below the output of a large plant by 1969, and absorption was less than half of output. The final column of the table shows that the entire non-Communist world could be served by no more than eight efficient-sized plants, conservatively defined. The production economy of scale barrier even into the world industry is therefore extremely formidable.

COSTS OF TRACTORS OF DIFFERENT SIZES

Tractor Costs' raw data do not permit a discounting approach to the estimation of production costs for each individual size of tractor, but we have Commission estimates of costs of the individual sizes at a 60,000 unit output. The cost of a 35 to 45 h.p. diesel tractor was given as $2,601 (with a gasoline engine at this volume the cost would be about $200 less); the 80 to 100 h.p. model, $3,465; and the 125 to 135 h.p. model, $4,682.[16]

It is significant that of the $2,081 rise in cost between the small and large models, an 80 percent increase, 86 percent of the difference is due either to increases in outside purchase costs (50 percent) or increases in foundry costs (36 percent). Further, of the $1,042 difference in outside purchase costs, $510, or nearly half, is for tires alone. The increased foundry costs result almost exclusively from the more or less uniform scaling-up of part sizes. Stamping, machining, assembling, and support costs are each only slightly higher.[17]

These estimates illustrate the point that a small tractor differs from a large one in little else but physical size. Nevertheless, lower unit costs for any one size of machine and greater economies of scale would undoubtedly accompany complete specialization in one power size only and would also probably be non-negligible if the three different sizes produced were closer together, thus permitting a greater commonality of parts use.[18]

Table 3-3. Production and Domestic Sales Divided by 90,000 Unit Plant

	U.S.		Canada		U.K.		France		Germany		Italy		Australia		Total Non-Communist Sales
	P	S	P	S	P	S	P	S	P	S	P	S	P	S	
1950	5.5	4.8	.1	.6	1.3	.4	.1	.2	.6	.4	—	—	—	.2	7.9
1955	3.7	3.0	—	.3	1.5	.5	.7	.7	1.6	1.2	.2	.2	—	.2	8.3
1960	1.7	2.0	—	.3	2.0	.5	.7	.8	1.3	1.0	.4b	.3b	—	.2	6.9
1965	2.7	2.2	—	.3	2.0	.5	.7	.8	1.2	.9	.6b	.4b	—	.2	7.5
1969	2.2	2.2	—	.2	2.0	.5a	.8	.9	1.1	.8	.7	.4	—	.2	7.8

a1968.
b1966.

Source: See Figures 2-1, 11-1, and 11-2.

In the 1968 North American tractor market, not three, but as many as six or seven power choices (although bounded in horsepower by those actually investigated by the RCFM) were actually produced by most manufacturers. In addition, there were a number of other major options (see the following chapter). How much would such offerings make actual operating costs from new plant differ from those estimated by the RCFM? *Tractor Costs* does not ignore this problem. It allows for the cost of assembly and warehouse space for a usual range of add-drop optional equipment, and it is admitted that the actual complications on the assembly line add slightly to unit cost. More important are engine and transmission options which would "increase the man-hours and machine time devoted to set-up changes and also increase inventory requirements."[19] Although the *Report* gives a more vivid picture than does *Tractor Costs* of the considerations involved, the only conclusion is that "while it is clear that the trend towards additional variety in terms of sizes, options, and models has added to farm machinery costs (most of the preceding discussion had concerned itself with tractors or combines), it is not possible to quantify these added costs in any simple way. In some instances, at least, the additional features have been provided at modest additional cost."[20] This conclusion appears to be a reasonable one considering the very large part of the variation in product under consideration which involves physical size and power. Many parts, sometimes including large castings and sub-assemblies, would have been appropriate for machines of different sizes and horsepower ratings. To the extent that this was the case, scale economies would not have been affected by a more "dense" line than actually studied. Furthermore, engine blocks, especially designed to accommodate differing numbers of cylinders and varying displacements, were a source of commonality not seriously considered by the RCFM, although such engines were actually being built by at least one major manufacturer (Ford) by the time the study was conducted.

CHANGES IN COSTS WITH TEMPORARY FLUCTUATIONS IN VOLUME

The RCFM used actual industry experience in an investigation of changes in unit costs from fluctuations around designed plant capacity. These estimates suggest that at the 20,000 unit level a 20 percent drop in production brings a 7.5 percent increase in unit costs, while a 20 percent expansion brings a 3.8 percent saving. For 60,000 units the figures are 7.1 percent and 5.5 percent, and for 90,000 units, 7.9 percent and 3.8 percent.[21]

Attempting to distinguish between "fixed" and "variable" costs is a difficult task because some of the costs which a textbook might treat as variable are in fact not likely to be completely avoided when demand drops temporarily, nor are the inputs which these costs reflect in fact augmentable at the same price when demand temporarily increases. The most important causes of the former phenomenon are contractural obligations and the desire to conserve trained manpower, and, of the latter, the necessity to make overtime payments. These considerations are reflected in the figures just presented. It should also be pointed out that however "variable" one treats various categories of manpower expense, at all volume ranges considered, material costs account for over 70 percent and capital costs no more than 15 percent of unit costs. This leaves only about 15 percent of unit costs to be puzzled over. Together with the level of concentration, these figures would make "cut-throat competition" during depressed demand periods most unlikely [22] in most markets at any time, despite the fact that cross-elasticity among established sellers is probably very high, as suggested in the following chapter.

THE COMMISSION'S BRITISH COST ESTIMATES

There have been no studies of the costs or economies of scale in tractor manufacture for countries other than the United States (and Canada). The Royal Commission did, however, make an attempt to apply part of its procedures to British conditions in the *Special Report on Prices.* [23] Labor productivity was assumed to be equal in the U.K. and the U.S., an assumption based solely on a suggestion from the manager of Ford's Basildon operation that productivity in the new plant was comparable with that achieved in the U.S. [24] Many less justifiable assumptions were made. For instance, capital costs were assumed to be the same in the U.K. and the U.S., and purchased parts were uniformly estimated to be 25 percent cheaper in the U.K. following the 1967 devaluation of the pound.

Given the weaknesses of the estimates made by the Commission, it is not surprising that, in the last analysis, it placed its confidence in an independent industry-source estimate of $1,380 (U.S.) for a 40 h.p. diesel tractor (immediately following the 14 percent devaluation of 1967). The source of the estimate is not given, but the figure was "not challenged by farm machinery companies manufacturing in Britain with whom the Commission discussed the question." [25]

Although the lower British costs, by comparison with the U.S. paper plant estimates for a small tractor, may be due in part to a narrower range of horsepower sizes as well as lower input costs and greater overall volume, the actual number of models produced in Britain by both Ford and Massey-Ferguson at the time of the estimate was four, and a complete range of options was available. There seems to be no reason not to accept tentatively the cost estimate for the small British tractor, and to use it with caution as a benchmark from which to estimate other British tractor costs. If the same economies of scale could be expected in percentage terms for the U.S. and the U.K. and costs for the larger U.K. tractors are calculated by taking the same relative costs as were found in the U.S., the relations between costs in the two countries would be as shown in Table 3–4.

The estimated cost differences are enormous; the British figures are over 35 percent lower. The Commission's decision to treat British costs at that time as no more than 75 percent of those in North America for comparable volumes appears to be conservative.

OTHER EVIDENCE ON INTERNATIONAL COST DIFFERENCES

There is corroborative evidence for lower British costs earlier in the post-war period. Dunning estimated British costs to be lower in 1953 or 1954, and in 1954, *Implement and Tractor* reported that "Tractors are currently being produced at lower costs, especially in the sterling block nations, especially as pertains to wage rates."[26] Albert Thornborough, later head of Massey-Ferguson, led a study team which in 1954 reported that a soon-to-be produced Ferguson tractor, the TE-35, could be produced in Britain and delivered to Detroit at a savings of over $200.[27] A production cost difference of $300 would be a conservative estimate, as transportation costs were almost certainly more than $100 at the time. When the TE-35 was actually introduced in the United States from U.S. plants (there were contractual obligations with U.S. suppliers which made this necessary), the suggested retail price was $1,845. Costs of goods sold

Table 3–4. Estimates of U.S. and British Tractor Production Costs—1968 (U.S. Dollars)

	Annual Unit Output		
	20,000	*60,000*	*90,000*
U.S. Small	$2971	$2653	$2441
U.S. Medium	$3958	$3534	$3251
U.S. Large	$5349	$4776	$4393
U.S. "Average"	$3801	$3381	$3094
U.K. Small	$1892	$1690	$1555
U.K. Medium	$2518	$2248	$2068
U.K. Large	$3399	$3035	$2792
U.K. "Average"	$2379	$2124	$1954

Note: U.K. small tractor costs at 90,000 units per year were first estimated by eliminating the supposed savings at 120,000 from the independent estimate of cost at that level. $1,380 + $150 in engine saving + $25 in modular construction saving = $1,555. U.S. average tractor cost found by discounting at the 8 percent rate has been corrected for the outside purchase part overestimate discussed in a previous footnote. There is no way to rework the Commission data to find costs for individual tractor sizes by discounting over the life of the project; *Tractor Costs'* estimates of relative costs are used. For additional assumptions, see text.

in the farm equipment industry were running between 50 and 55 percent of suggested retail prices at the time, and tractors were regarded as a relatively high mark-up item. If tractors sold for about twice their accounted manufacturing cost and the model 35 was a typical machine, then the British costs were nearly one-third lower than those in North America. In 1954, the output of Ferguson machines in Britain exceeded 60,000, while the combined outputs of the Massey-Harris and Ferguson tractors from U.S. plants was only about 25,000, but, while scale economies must have been of some importance, the Thornborough report did not discuss them.

Another significant piece of evidence relates to British costs in 1960. International Harvester estimated that the B-275 could be built for $500 less in England than in the U.S. and further affirmed that: "Some of the components such as tires cost more in the European market; factories are not as automatic and more man hours are required to build each tractor, but the big saving is in total labor cost for each completed unit because of the lower wage scale."[28] The B-275 was actually sourced from Britain, but it was priced in the U.S. to meet competition there and very much higher than in the U.K. If the prevailing price in the U.S. were approximately twice representative U.S. production costs, then the savings represented over 35 percent of U.S. costs. Furthermore, the volume of the U.K. factories of International Harvester was probably no greater than 30,000 at the time while U.S. output was over 50,000 and had probably been nearly 90,000 in the previous year, so scale economies clearly favored U.S., not British, sourcing.

The only official recognition of the differences in tractor production costs between the U.S. and the U.K. known to the writer came in the House Committee on Agriculture's "Cost-Price Squeeze" Hearings of 1961. Under questioning, a vice-president of Massey-Ferguson estimated that a tractor could be produced in Britain for 20 to 25 percent less than a virtually identical machine in the United States.[29] Economies of scale were not mentioned as an important factor in the cost difference, although British output was over 60,000 and U.S. output under 30,000. The clear implication of the Massey-Ferguson statement was that wage rates were the principal source of difference, and later testimony by a representative of the United Automobile Workers (UAW) did not dispute this explanation.[30]

A 1964 dissertation in the field of marketing examines the loss of overseas markets by the U.S. tractor industry. Most of Linneman's conclusions about costs are inferred from price differences, and although these may be broadly indicative, there is evidence of international price discrimination in the industry, and thus price comparisons warrant separate discussion. Linneman's industry sources did agree, however, that the U.S. could not meet British and German competition in the French market in the fifties, even without discriminatory currency controls against the dollar area.[31] Further, both German and British production were carried out at significantly lower cost than that of French production,[32] at least until the large devaluations of 1957 and 1958 (which amounted to 29 percent). British costs were much lower than those in Germany at this time as well, but the scale of output was much larger for the major firms in the U.K. than in Germany, while the tractor volume of most German firms was comparable with those in France.

A 1958 government report on the Australian industry cited by Linneman blames high unit cost resulting from a low volume of production (something over 2,000 units in 1958) and higher costs of Australian produced purchased components for the failure of Australian-produced tractors in the 30 to 40 h.p. class to compete successfully with similar British-produced machines despite a subsidy of up to 22 percent of the wholesale price. An increase in the subsidy schedule (based on percentage of local manufacture) to a maximum of 37 percent rendered Australian machines at least temporarily competitive. Wage rate differences between Australia and the U.K. were not thought to be of substantial importance in the disadvantage.[33]

An RCFM study by its Director of Research, Neil B. MacDonald, attempted to compare the United States and Canada as locations for the production of farm machinery in 1968. Labor costs (including differences in productivity and fringe benefits) in Winnipeg were only 61 percent of those in Brantford, Ontario, and only 52 percent of those in Moline, Illinois.[34] This helps to explain the great competitive advantage of Winnipeg-based Versatile, a small producer of four-wheeled drive tractors, combines, and implements (see Chapter Seven). MacDonald's study concludes, however, that the Winnipeg advantage might well be lost if the firm were to become larger and attract the attention of organized labor.[35]

Neglecting movements in the U.S.-Canadian exchange rate, a few currency realignments among countries in the study did take place after the 31 percent devaluation of the British (and Australian) pound in 1949. In addition to the 14 percent further devaluation of the pound in 1967 and the 29 percent French devaluations of the late fifties already mentioned, there were the 5 and 9 percent revaluations of the German mark in 1961 and 1969, and the further devaluation of the franc by 11 percent in 1969. Our study largely ends before the massive currency realignments which began in late 1971.

Clearly the pattern of national costs at any given time and its rearrangement through currency valuation changes are important structural characteristics which ought properly to loom large for predictive and evaluative purposes. Unfortunately our direct information is exhausted by the material just reviewed. Specifically, one would predict that in addition to the realization of scale economies through maximum feasible consolidation of manufacture (to be discussed in Chapters Seven and Eleven), there would be a strong tendency toward secular adjustment both within firms and among them toward manufacturing locations with low "absolute" costs—tariff barriers permitting (see Chapter Six). In particular, one would predict that the U.K. should loom larger as a manufacturing base than as an absorber of tractors, and Table 3-3 confirms that this has in fact been the case. It will be established in Chapter Ten that Italy too has been a low absolute cost producer, and it rapidly expanded export markets after the home market grew to significant size. Normatively, as was suggested in Chapter One and will be further argued in Chapter Twelve, "good" economic performance by the world industry should be partly measured by the extent to which its output comes from areas of low sustained absolute cost.

ECONOMIES OF COMPLEMENTARY PRODUCTION

Complementary production economies can be defined as those savings which are attributable to extra volume of production of certain parts because of their use in products other than that of the industry being examined. One obvious place where such economies can be achieved is in engine production which accounts for about one-quarter of all tractor costs. In addition to the economies in tractor manufacture achieved between 20,000 and 90,000 units per annum, it has been estimated that economies of scale in engine manufacture may obtain up to between 260,000 and 280,000 units, at least in the United States,[36] and no existing tractor maker could achieve such an output for tractors alone. Nevertheless, most of the world's major automobile manufacturers have engine outputs in this range or higher.

The situation is complicated by the fact that somewhat differing engines are used for various motor vehicles. Nevertheless, there is substantial similarity, and European tractors have often been powered by only slightly modified car or truck engines.[37] A firm which produced construction machinery could also gain some complementarity with wheeled tractors if it produced small crawlers, because it would be able to share some transmission parts, stampings, and hydraulic components.

What of production complementarities for full-line farm machinery firms? The combine is the only other major piece of self-propelled farm equipment, and it is produced in much smaller volumes than wheeled tractors and with a somewhat different engine. In 1966 combine production was only 15 percent of tractor production on a unit basis, worldwide.[38] Engines excepted, the manufacture of tractors has very little in common with that of most of the other machines sold as farm machinery (including combines), and virtually all tractors are produced year-round in separate plants.[39] There is thus little production complementarity between tractors and other farm equipment except perhaps in some minor use of foundry output.

Economies of scale are still possible in purchasing. If, however, as Schwartzman suggests,[40] such economies relate almost exclusively to the buying of highly fabricated products such as generators or windshield glass, a firm making construction equipment, for example, would, comparing a greater with a lower volume of final sales, be gaining greater such economies than a producer of a range of farm equipment. After the combine, the degree of complexity of farm machines falls off sharply; such machines as swathers, hay-balers, and wagons are far simpler in design. Thus, there would seem to be only minor economies to tractor production from the manufacture of a full-line of farm equipment.

POST-PLANT COSTS AND ECONOMIES OF SCALE

In *Oligopoly in the Farm Machinery Industry* Schwartzman makes an attempt to estimate post-plant economies of scale in North America.[41] He combines his own educated guesses with time series and cross-section industry data available to the Commission and uses some expense estimates from the *Tractor Costs* study. He determines that a firm producing 20,000 tractors a year would gain slightly greater percentage cost savings in general and administrative expenses, branch warehouse operation, inventory and dealer finance, and research and development taken together by moving to the 90,000 unit level than it would gain in production cost reduction (see Table 3–5.) The unusual industry practice of completely financing dealer inventories will be discussed in Chapter Five. Schwartzman's analysis implies that these additional expenses move average tractor cost at the 20,000 level from $3,801 [42] to $5,066 or an increase of 33 percent and from $3,094 to $4,004 or an increase of 29 percent at the 90,000 unit level. The combined production and distribution unit cost reduction is $1,062 or 21 percent between 20,000 and 90,000 units.

Although one might quibble with many of the assumptions made by Schwartzman, his greatest conceptual failing seems to lie in completely ignoring product mix and estimating economies solely on the basis of dollar volume. An examination of his data, however, doesn't indicate that in most cost categories there would be a major cost difference between a tractor-maker and a full-line firm of similar dollar sales. It is possible that the costs of operation of regional wholesale outlets at any level of sales would be slightly higher for the tractor-maker than for the full-line firm because of the economies which the latter would realize from dealing with fewer retail outlets, but it is also possible that the estimates

Table 3–5. Non-Plant Economies of Scale in North America (Non-plant expenses as a percentage of total sales)

	Annual Unit Output		
	20,000	*90,000*	*Difference*
General and administrative	4%	3%	1%
Branch warehouse	9	7	2
Inventory financing	7	5	2
R & D	5	3	2
Total	25%	18%	7%

Source: Derived from Schwartzman, *Oligopoly*, pp. 67–70.

of both the volume of expenditure and economies of scale for research and development in tractors, by contrast with farm machinery in general, might be excessive. All things considered, we are probably safe in assuming, as does the Commissioner in the *Report,* that in North America, "the behavior of total costs per unit as volume increases does not differ significantly from costs at the manufacturing level." [43]

For purposes of evaluating entry barriers, however, it must be stressed that a full-line firm (with no other activities), at any given level of tractor sales, will be between two and three times as large as a firm making tractors only. This would provide a cost advantage over the range studied of perhaps 2 to 3 percent at any given level of tractor sales. It is assumed that there is no difference in cost of inventory financing for firms selling an equal number of tractors whether they are full-line or not. The economies of scale in that category of expense turn almost entirely on the average size of dealer served and his turnover (larger dealers have higher turnover), and it is not unreasonable to assume that an independent tractor manufacturer and a full-line firm would sell the same volume of tractors per dealership at comparable tractor outputs.

It is obvious that the economies of scale estimated for distribution do not have the same meaning as estimated production economies; this is true for two reasons. First, such economies are only roughly inferred from actual costs incurred, rather than estimated independently as was the case for production economies in the "paper plant" study. Second, while meaning attaches to "best practice technique" in the production of tractors, economies in distribution are intrinsically ambiguous without a precise specification of what "the product" is and an examination of alternative ways of accomplishing the same end. In fact, the estimates reflect credit practices and a supervised retail outlet system which, as Chapters Five and Eight will argue, contain a very substantial promotional component. They would nonetheless probably have to be accepted as a given by any potential entrant, and so are of great interest.

Until the 1960s, dealer inventories were not usually financed by the manufacturer outside of North America. During that decade, however, for reasons yet to be explored, this practice grew substantially in most markets, although the provision of credit never became as lavish as in the U.S. and Canada. Furthermore, except for Australia, the rate of sales per dealer and presumably inventory turnover are higher in other study markets than in North America, so the funds tied up in distribution would be lower even if credit were provided equally in North America and elsewhere. Finally, because of the greater geographic compression of non-North American markets [44] (again, except for Australia) other distribution expenses (such as the operation of regional wholesale outlets) are lower than in the U.S. and Canada. The first two cost categories are of a discretionary, promotional nature and are much in excess of what would be expended if dealers themselves paid for their own inventories (dealer expenditure on inventories could still be in part competitive and therefore not completely necessary from a social point of view); the competitive forces which induced them outside of North America will be discussed in Chapter Eight. On the other hand, geographical dispersion must at least somewhat increase cost under any distributional regime. For all three reasons, however, both levels of firm cost at any scale and economies of scale in distribution must have been considerably lower elsewhere than in the U.S., Canada, and Australia even at the end of the period.

Whatever levels of expenditure and economies of scale might prevail in research and development in North American tractor-making, they are probably lower abroad, because machines produced there are usually less original in design than those introduced into the North American market; this point will be further developed in Chapters Seven and Ten.

CAPITAL REQUIREMENTS ESTIMATES

The best source of information on capital requirements is, once again, *Tractor Costs* of the Royal Commission. Table 3-6 shows its estimates of manufacturing capital costs. Initial capital requirements in North America rise from $74,657,000 (U.S.) for 20,000 units to $259,109,000 (U.S.) for 90,000 units.[45] The 90,000 unit plant cost estimate by 1968 more than doubled Bain's estimate of $125 million dollars [46] in the early fifties which was sufficient to place tractors in the highest capital requirement category. Although European costs might be considerably lower, the order of magnitude of the funds involved for efficient production would still be very great by almost any standard.

To begin de novo with a distribution system similar to the one which prevailed throughout in North America would add enormously to capital requirements. Wholesale distribution assets of inventory and accounts receivable from dealers must be added (even if final farmer finance, the importance of which remains to be established, is ignored). One can infer from Commission material [47] additional capital requirements varying from $67,440,000 at 20,000 [48] to $303,480,000 at 90,000. These figures do not include any capital investment

Table 3-6. Capital Requirements for North American Production and Distribution (U.S. Dollars) 1968

	Annual Unit Output		
	20,000	*60,000*	*90,000*
Manufacturing	$ 74,657,000	$171,689,000	$259,109,000
Distribution	67,440,000	202,320,000	303,480,000
Total	$142,097,000	$374,008,000	$562,589,000

Source: See text and footnotes.

in branch and central office buildings and furnishing or in repair parts inventories.

Again, until at least the mid-sixties, selling outside of North America would have necessitated much lower levels of working capital, because of the greater density of the market (again, excepting Australia). The lack of necessity of dealer finance, which in North America accounted for over half the estimated distribution capital in use [49] apparently did not substitute significantly for company inventory; thus it largely represented wasteful promotional expenditure. Nevertheless, for reasons which will be discussed in Chapters Seven and Eight, the absolute cost barrier presented by the growth of dealer financing would appear to be substantial and increasing in most non-North American markets. The option of using an independent wholesaler will be explored, but it is one which has increasingly proved to be undesirable.

No other barriers to entry of the kind usually included in the "absolute cost" category were found either by Bain or by the RCFM for the North American industry, and none appear to prevail elsewhere.[50]

Chapter Four

The Product: Development, Differentiability, and Differentiation

This chapter has three purposes. The first is to trace briefly the technical development of the tractor and to show the extent to which machines sold around the world as tractors became increasingly alike in the post-war period. Secondly, an attempt will be made to evaluate those characteristics of tractor design, use, and conditions of sale which contribute to the ability of manufacturers to achieve brand preference. The third purpose is to begin an evaluation of product differentiation as a barrier to entry into the tractor industry; a pivotal differentiation barrier, established distribution systems, will be treated in the following chapter.

PRE-WAR TRACTOR DEVELOPMENT [1]

Most of the technical development of the tractor had already taken place by World War II, and it is not surprising that much of it occurred in the United States which dwarfed the rest of the world in pre-war production and use. Although the first successful gasoline tractor was built in the United States in 1892, there were only about 10,000 tractors on U.S. farms by 1910. By the end of the next decade the number had grown to nearly a quarter million, in large part due to the acceptance which greated Henry Ford's "Fordson" of 1917.[2] The machine was light in relation to its power and low in price, and it therefore provided a real alternative to animal power for many farmers. The Fordson accounted for 25 percent of the 135,000 units sold in 1918 and 75 percent of the 158,000 1925 market. The casual observer would easily recognize the Fordson as a rather austere prototype of the modern machine, and the features it lacked could usually be found elsewhere. As two American experts, Dieffenbach and Gray, have concluded, "Tractors in 1920, considered collectively, embodied fundamental principles of design that exist, perhaps in refined form, in today's tractors."[3]

Despite its great success, the Fordson was far from an ideal machine. When plowing, resistance to the trailed share from a log or rock would pull the tractor back on its rear wheels, throwing the driver from his seat and sometimes pinning him under the machine. Further, the clearance and wheel spacing of the Fordson limited it to such uses as plowing and driving threshing machines; it was not suitable for work in a growing field. International Harvester's aptly named "Farmall," introduced in 1925, was both much better balanced than the Fordson and designed for all types of field work. A far greater number of farmers could now completely replace horses with mechanical power and therefore justify the tractor's acquisition cost. The success of the Farmall and several imitators caused Ford to withdraw completely from the North American tractor market in 1928.

Between 1920 and the late thirties, there were several other significant developments. In 1923, John Deere introduced the Model D which was to be produced with only slight modification until 1960. Like the Fordson, it was not a row-crop machine but it was low in price, yet powerful, and could run on almost "any fuel that could be poured into the tank."[4]

Somewhat similar in appeal to the Deere tractor, but smaller, was the German machine developed by Heinrich Lanz of Mannheim. The Lanz Bulldog, introduced in 1921, had only one cylinder (Deere's had two, and most other U.S. machines at the time were four-cylinder), could run "on any type of gasoline or vegetable oil,"[5] had a minimum of moving parts, was low in price, and quickly became the leading tractor outside of North America. The basic design was still being produced in models of various power when Deere took over the company in 1956.[6]

During the thirties considerable progress was made in improving the balance of small tractors while reducing their production costs and price. The leader in these developments was Allis-Chalmers which also successfully experimented with the use of rubber tires for tractors. This greatly improved effective horsepower, allowed the tractor to be moved across the growing number of paved roads of the U.S. and Canada, and immensely improved driver comfort. Although only 14 percent of all new U.S. tractors were mounted on rubber in 1935, the number reached 85 percent by 1940 and nearly 100 percent by 1950 where it remained. Almost all European production in the post-war period has been rubber-tired.

The truly revolutionary development of the pre-war period, the "Ferguson system," came not from the United States but from the British Isles, and was introduced into North America in 1939 through a production arrangment with the Ford Company which lasted until 1946, after which Ferguson was forced to find another supplier, and Ford continued to sell a very similar tractor (see Chapter

Seven). The machine was the culmination of years of in-field experimentation by the Ulsterman, Harry Ferguson.

The Ferguson system with a special hitch for automatic hydraulic control of integrally-mounted implements is probably the only identifiable "quantum jump" in the otherwise evolutionary technical development of the tractor, a development which has relied heavily on adapting improvements in technology made in the automobile industry and the firms supplying it.

Simply stated, the Ferguson system involved two related components: the "three-point hitch" and a hydraulic control system. An ordinary tractor pulling against heavy resistance as in plowing, encounters engine torque tending to lift the front of the tractor, sometimes causing it to tip over backwards. The conventional solution was to make the tractor heavy—sometimes by adding weights to the front end. By contrast, Ferguson's "draft control" system used the tendency of an implement to rotate between two hitch points located low on the rear of the tractor against the third "top link" to activate a hydraulic mechanism which in turn tended to lift the implement, adding weight and traction through the rear wheels. The system allowed the implement to adjust to changing soil resistance as it actuated changing pressure on the top link. The top link also served as a brace which prevented the lifting of the front end of the tractor. A short, light tractor could thus be more stable and have greater traction than was previously possible.[7]

This system, embodying refinements contributed by various manufacturers, came to be used almost universally for light tractors in North America and the United Kingdom by the late fifties, and most Continental makers copied the system within the same period. Further experimentation showed that Ferguson principles could be applied to trailed implements, and the three-point hitch and associated hydraulics became almost completely universal. Only a few of the largest North American machines, designed especially to drag gangs of trailed implements, omitted some form of the system in the late sixties.

POST-WAR TRACTOR DESIGN

In the post-war period there have been some minor developments in tractor design in addition to the diffusion and improvement of the Ferguson system. Power steering was added to tractors with little necessary adaption by purchasing it in the first instance from outside suppliers. The demand for such power assistance arose in North America from the popularity of front-mounted tillage equipment on early post-war machines and from their increasing size; both factors made unassisted steering onerous.[8]

Another automotive feature adapted to tractor use was the automatic transmission, which by the sixties was available for nearly all tractors sold everywhere and had become standard equipment on many tractors sold in North

America, especially the larger units. Increasingly sophisticated shift-systems were offered over the period, but most tractors with automatic transmissions both in North America and elsewhere were by 1970 fitted with rather simple units, scarcely differing in price from those on automobiles.[9]

Chapter Two reviewed the differences in fuel type and size among the different markets in the study at the war's end and their subsequent developments. Over the entire period there has been a substantial North American demand for machines larger than sold elsewhere in any significant number. Larger engines and turbocharging were obvious answers to the North American challenges of increasing field speed and propelling ever-larger capacity implements. Less easy to deal with was the impact of increased tractor weight on the load-bearing properties of soils, and new solutions in tractor configuration were sought. The simplest modification to overcome the problem, double rear tires, merely lengthens the axle to allow for an extra tire on each side. A more elaborate solution was found in four-wheel drive tractors which have treaded and often large front wheels to assist with traction and load distribution. The most popular units of the latter kind in the U.S., Canada, and Britain have all four wheels of the same size and employ variously, front wheel, back wheel, four-wheel and pivoted-frame steering.[10] These machines undoubtedly cost more for an established tractor maker to produce than a similarly powerful but "scaled up" version of an existing model, but the underlying designs are just as simple as for a standard tractor. By 1970, only a few percent of total units were thus constructed, but the percentage will undoubtedly grow as horsepower continues to rise. Unconventional designs of various kinds amount to only a few percent of the market in Germany and Italy [11] and are even less important in other study countries. Further, their importance does not appear to be increasing in any of the non-North American markets.

Even when the basic configuration of the tractor is fixed, there are still minor possible variations. Both front and rear wheels may be adjustable (the rear ones sometimes by hydraulic power), the front of the tractor may be fitted with single, or more often closely-spaced double, wheels for a tricycle effect (but these machines have proven unstable, and the tricycle's main objective, allowing the front wheels to go between rows, can be accomplished with an adjustable front axle). The frame may be built high (hi-clearance) for cultivating some crops, made narrow for vineyard work, or set low and put on large soft tires for work on soft ground and hillsides. Nearly all of these features entail very minor variation in tractor design, and in addition to size, engine-type, and several other minor add-drop options, they nearly exhausted design variation after the mid- to late fifties.[12]

An overall consideration of the world-wide tractor design in the post-war period conclusively shows that tractors became increasingly similar. At the war's end and for a decade thereafter, the Ferguson design was unique or only partially imitated. Continental manufacturers were experimenting with various design configurations and engine types and there were minor but noticeable differences in the thrust of the offerings of the "old line" U.S. producers. While some experimented with large six-cylinder power units, Deere continued to market several sizes of its dependable two-cylinder design. A look at the offerings of leading firms by the early to mid-sixties, however, reveals a striking convergence. Depending on markets served and marketing strategy, the largest offering varied considerably (up to 120 h.p. for a standard design two-wheel drive model), but otherwise features were remarkably similar from configuration to engine design [13] to hydraulic system, and the larger standard units within a line differed little except in size from the smaller ones, as evidenced by increasing intra-line parts interchangability.[14]

There have been few important patents in the tractor industry;[15] the outstanding exception was the set of patents surrounding the Ferguson system, but these were developed around and improved upon very successfully by Ford and several other manufacturers.[16] Indeed, the Ferguson case illustrates the general pattern of industry innovation. It was not for want of unknown or unattainable techniques that several manufacturers waited for more than a decade to copy Ferguson [17] but from a misreading of market trends. The introduction of power-steering, higher horsepower, four-wheel drive and other rather obvious developments did indeed shift market shares, but there is little in any of the innovations which any firm in the industry could not easily have been the first to introduce if it had chosen to do so.

Lawrence White's recent study of the automobile industry notes,

> Perhaps the most striking thing about automotive technology in the postwar period has been the lack of fundamental change or advance. Cars built in 1968 are not fundamentally different from those built in 1946. . . . Even in the areas in which modern cars do differ from their early post-war predecessors the basic technology had been developed before the war, and post-war developments represented achievements in refining this technology rather than in any fundamental change.[18]

This seems to be equally true of tractor development, and hence the truth is also more obvious, because there is less stamped sheet metal to disguise reality.

PRODUCT DIFFERENTIABILITY

The extent of "differentiability" in tractor manufacture must be assessed in light of the above material. Differentiability depends upon product traits which "make it likely that a producer can create the illusion in the buyer's mind that his brand possesses certain virtues that cannot be duplicated and that are worth a slightly higher price."[19] Specifically, the considerations are: 1) whether the physical qualities of a product are immediately obvious to a buyer or are either difficult to assess or capable of assessment only over a prolonged period; 2) whether the price is high enough for the purchase to warrant scrutiny or 3) often enough for the market alternatives to be well known to the buyer; and 4) the degree to which social esteem rather than private usefulness helps sway the purchaser's decision.

Only certain dimensions of tractor performance are defined, measured, and available to a prospective purchaser. Since 1920 the state of Nebraska has measured certain dimensions of performance as a condition for machines to be sold in the state.[20] Testing has been devoted almost exclusively to engine power, turning radius, and fuel consumption, although in 1970 noise level testing was added.[21] The growth of world trade in the late fifties brought to the attention of the Organization for Economic Cooperation and Development (OECD) the usefulness of a standard test which could increase machine comparability internationally and avoid costly testing duplication. In 1959 the "OECD Standard Code for the Official Testing of Agricultural Tractors" (revised in 1965 and 1970) [22] was introduced along lines very similar to the original Nebraska tests. All OECD countries producing tractors subsequently adopted the code and began official testing except for the U.S. which continued to rely on the Nebraska procedures. The Code provides standard procedures for measuring power, turning, balance, braking, noise, and hydraulic lift capacity. In spite of this testing, however, there is no information available to purchasers on relative performance over time, and as a tractor may have a useful life of up to 15 or more years, product reputation retains great importance, as it can provide assurance about durability, adequate parts provision, and resale value.

Certainly the purchase price—which in North America ranges from that of a subcompact car up through that of the largest domestically produced luxury sedan—would tend to make the purchaser as careful and judicious as available information allows, despite the fact that tractors are typically purchased only every several years.

Although the tractor is a producer durable, it is nonetheless highly conspicuous to family and neighbors. The extent to which styling is regarded as important in the industry will be discussed in Chapter Ten, as part of a general discussion of product competition, but given the rather sharp confines in terms of basic deisgn within which styling changes would have to be developed and the rather small surface area amenable to manipulation, very much less importance would presumably attach to styling in tractors than, for example, in automobiles. Not only is there less to work with, but it is much more difficult to imagine a successful appeal to basic hopes and fears through a piece of agricultural machinery.

Overall, it would seem that the differentiability of tractors, considering only the factors mentioned previously, would be quite high, although, as increasing agreement about appropriate design was achieved, the ease of physical imitation and rather static technology dictated that differentiability perforce became increasingly focused on performance over time, i.e., dependability and durability.

PRODUCT DIFFERENTIATION AS A BARRIER TO ENTRY

When Bain examined the product differentiation barrier to entry into the U.S. industry providing "tractors and large, complicated farm machinery," he placed it in the highest category [23] among his more than twenty industries and attributes the barrier to "brand allegiances based on product reputation, customer service, [and] established dealer systems."[24]

Performance over time is a function of the ability of the manufacturer to provide a product of given durability with spare parts back-up when necessary, and the (almost always independent) dealer's ability to do whatever work is necessary, with skill and dispatch. Suppose the latter problem were to be solved and that the farmer could be persuaded of the initial performance of a new brand; he would still be worried about the resale value of the machine and the speed of availability of spare parts. The differentiation advantage of established sellers would remain high; as Bain suggests, when combined with established dealerships, they become extraordinarily high, at least in North America. On the other hand, if manufacturers could establish similar reputations and dealership systems, one would expect to see similar prices for established firms as well as much lower prices or very high promotional expenditure by newcomers.

The following chapter will discuss conditions of distribution in detail, but before proceeding, one final point of a priori theorizing about "product reputation" seems appropriate. Chapter One mentioned marketing "spillover," and this factor may be important in the tractor industry in a rather special way. It seems reasonable that there has been and continues to be a considerable asymmetry between the advantage of being a leading seller in North America for sales outside that market and the acceptance in the U.S. and Canada of established names elsewhere for which the differing national levels of agricultural productivity rather than firm-specific factors are in part responsible. The aura of productivity which attaches to American agriculture may serve as an entree elsewhere for firms established as suppliers to that market. On the other hand, it is also possible that producers with a home base in a country with a poor international image in agriculture might suffer from it. Some partial confirmation of these hypotheses will be presented in Chapters Seven through Eleven.

Chapter Five

Distribution Systems

The development of an effective distribution system is of paramount importance to any tractor maker. Especially careful attention to North American distribution is appropriate for several reasons. First, the United States is by far the largest single market in the study, and, with Canada, accounts for nearly half of all sales in the seven countries, and distribution networks appear to have provided a formidable barrier into the North American market. Second, because North American-based firms could be expected to attempt to initiate marketing practices proven at home when they move abroad, their importance as barriers to entry is of great interest for the world industry. Finally, by contrast with the situation in most other markets, the distribution practices of the major firms in North America were already well established before the war and changed little thereafter; they therefore deserve treatment as structural parameters.

NORTH AMERICAN DISTRIBUTION

The Wholesale-Retail Relation
Wholesale distribution in the United States and Canada has traditionally been handled by the manufacturer through "branch houses" which serve as offices for regional company sales personnel and warehouses for goods and parts between the factory and the closely supervised franchised retail dealer. The pattern has been subject to exception; some independent and cooperative wholesaling and retailing have been employed, and they will be discussed in Chapters Seven and Eight, but manufacturer wholesaling was already dominant before World War II and remained essentially unchanged for most firms in the post-war period.

Tractor dealerships in North America are independent and limited in number. Both of these characteristics require an explanation. White has suggested three possible a priori reasons why independent dealerships rather than factory outlets or company stores are employed in the automobile industry: 1) a smaller management personnel requirement, 2) a smaller amount of invested capital needed, and 3) a sharing of the risks of the automobile business.[1]

The first two of these apply with much less force for farm equipment dealerships (less than half the average annual sales of which are usually made up of tractors), and the third appears not to hold at all. Extremely expensive and elaborate manufacturer supervision of dealerships takes place, with one "blockman" or territorial representative for approximately 10 dealers; this overseer may even work on or close individual sales.[2] The possible saving to the manufacturer of management personnel is therefore considerably reduced. Further, it is the prevailing farm equipment industry practice to finance inventories of new and used goods in the hands of dealers, a practice known as "floor planning," which is almost completely unheard of in automobile retailing (except sometimes when dealerships are being started). A very large part of the capital cost of retail distribution is therefore borne by the manufacturing companies.

To test the third possible a priori reason for the existence of independent dealers, dilution of risk, White compares the standard deviation to average ratio of net profits on net worth for dealers and for manufacturers and finds the dealer ratio to be almost double.[3] Unfortunately, statistics are not available for the farm equipment industry which would make a similar test possible. During the fifties, Corporation Income Tax returns included farm equipment retailing with the sale of hardware; in the early sixties building materials were added, and only since 1963 have farm equipment dealerships been separately noted.

There are two other sources of profit information, but both appear to be biased in the direction of reporting high dealership returns. Dun and Bradstreet figures are available for a sample of firms for every year since 1963; unfortunately the survey is confined to firms with unrepresentatively high tangible net worth.[4] The most representative statistics available are probably those of the annual "Cost of Doing Business" Study conducted by the National Farm and Power Equipment Dealer's Association. One suspects that the dealerships voluntarily reporting to the trade association are the more profitable and better managed. Further, a definite upward bias is introduced by excluding returns from dealers who failed or otherwise discontinued business during the year.

Table 5-1 shows the returns of equity and assets (before tax) from the "Cost of Doing Business" Study together with the rate of return on equity (after tax) of Deere and Company after the end of the post-war boom through 1968. Deere was the only firm over the period which was overwhelmingly a farm machinery producer and also marketed by far the greatest part of its output in the U.S., and the stability of Deere's U.S. farm machinery profits was probably greater than that of other firms and perhaps even the industry as a whole (see Chapter Twelve).

Inspection of the table reveals that a simple coefficient of variation test for

Table 5-1. Rates of Return—Farm Equipment Dealerships and Deere

	Dealership Returns Before Tax as a Percentage of:		Deere After Tax Percentage Return on Equity
	Equity	Total Assets	
1954	7.4	5.8	6.7
1955	10.6	8.1	8.9
1956	11.9	9.1	6.1
1957	11.0	7.8	8.4
1958	13.2	7.9	13.4
1959	13.3	7.3	13.9
1960	10.6	6.5	5.0
1961	11.2	6.7	8.6
1962	14.5	7.9	9.3
1963	15.0	8.0	10.9
1964	15.8	8.1	12.5
1965	16.4	8.1	10.0
1966	19.5	8.4	12.8
1967	18.6	7.5	8.9
1968	15.7	5.4	6.5

Source: National Farm and Power Equipment Dealers Association, *Cost of Doing Business Study* (St. Louis: National Farm and Power Equipment Dealers Association), various years. Moody's *Industrial Manual*, various years.

risk, as limited as it may be in any event,[5] is clearly inappropriate here, because the dealership equity returns are markedly trended upward over time. Correcting this series for the trend,[6] the coefficient of variation remains over twice as high for Deere (.284) as for the dealerships on either the equity (.113) or asset (.129) measure. Though the data are far from ideal, it seems scarcely plausible that the farm equipment companies are shifting risk from themselves to their dealers. This result stands in contrast to White's conclusion from a number of differing comparisons that "the variability of returns faced by automobile dealers has been at least as great as the variability faced by automobile manufacturers and is probably even somewhat greater."[7]

All three of White's a priori reasons for the use of independent dealers appear to hold for the automobile industry, but their explanatory power in farm machinery seems weak or non-existent. There are further a priori reasons which White doesn't consider, however, which might be important in both industries. It may be that persons are willing to work longer hours and exert themselves more in an enterprise operated under their own name.[8] In the farm equipment business, moreover, there is another consideration which may be of greater importance, and it is undoubtedly important in the selling of cars as well. Dealership profitability depends to a large extent on how skillfully trade-ins are handled, and no company stores or cooperatives operated up to the early 1970s have found this problem handled with as much skill as when under the supervision of the owner himself.[9] Presumably some scheme could be created to share both profits and losses with a hired manager, but it remains to be devised, and to the extent that motivation is psychologically linked to an ongoing equity stake, this solution might still not be satisfactory. Overall strong dealer motivation is essential in light of the generally extremely high risk aversion of farmers concerning after-sales service.

There are important additional reasons why farm machinery dealerships are independent: there is a history of independent retailing going back to the days before the introduction of the tractor when most farm machines were small and were sold through outlets handling hardware and other merchandise. These trusted retail outlets became the main focus of manufacturer attention with the introduction of the tractor and the selling of full-lines. Dealers strongly favored a continuation of their independent status, and this preference had to be recognized because entrepreneurial talent on the countryside became increasingly scarce, and loyalty to established dealers probably increased with the increasing complexity of equipment.

Although comprehensive figures are not available, farm equipment retailing seems to be an occupation noted for passing from generation to generation. Deere and Company, the firm reputed to have the strongest dealer network, reported in 1965 that one-quarter of all its dealers were either the sons or grandsons of Deere dealers.[10] An important segment of the dealer group regards company attempts to intervene in the affairs of dealerships as a troubling sign of the decline of their autonomy, and would probably find becoming the absolute pawn of the manufacturer unthinkable.[11] Further, the company store, which usually operates in at least peripheral competition with established dealers, is viewed with a jaundiced eye,[12] and any company moving in the direction of substantially increasing forward integration would find itself with a very unhappy dealer group, perhaps part of which would find a more congenial place in another dealer system.

Even minor defection of dealers would be of great importance because there is an increasing scarcity of suitable people willing to live and work in the rural areas of the United States and Canada, where farm equipment (including four-fifths of all tractors) is sold. Figure 5-1 shows the distribution of tractors in North America. This manpower problem was noted even before the war,[13] and seems to have become increasingly acute since the mid-fifties. The trade press has been full of complaints about the companies' inability to find suitable replacements for men retiring from the trade and of dealers' difficulties in securing experienced service-department managers and even competent mechanics.[14] In recent years the major companies have begun to aid in the capital requirements for dealership facilities. This practice is one which has existed for many years in the automobile industry where the capital requirements are greater, but was apparently not thought necessary for farm equipment until the sixties.[15]

The foregoing suggests that each well-functioning dealership is an important company asset, and this is true not just because competent people are scarce but because customer loyalty is to a large extent dealer-oriented. One often observes a very different mix of equipment brands in adjacent areas of the Midwest with identical farming patterns and soil conditions despite the fact that all of the major brands of equipment are sold within each of the small adjacent areas, and dealers switching brands sometimes quickly regain their market share.[16] The key difference is dealer reputation, and this is what one would expect given the enormous costs to the farmer of slow or incompetent servicing and inadequate parts inventory; risk aversion with respect to dealers is understandably high. The relative importance of dealer as against brand loyalty has never been convincingly established; survey evidence is exiguous and sometimes contradictory. Nevertheless, most evidence suggests that *both* dealer and brand reputation are extremely important. One long-time industry observer believes that dealer service and the product line "image" are "on a par" with most buyers. There is, however, a rather small group of buyers who claim to be virtually indifferent to either brand *or* dealer (as long as minimum standards for the dealership are met) and will switch brand or dealer for only a few dollars' difference.[17]

Even if the problem of finding qualified dealers were less serious, the manufacturer would still want to limit franchises to some extent. It must be kept in mind that there are few rapidly growing markets in North America; new outlets for most manufacturers in most areas would therefore cut into the business of established dealers.[18]

Nonetheless, there are frequent situations in which no attempt is made to continue a dealership upon the owner's demise or an ongoing business may even be encouraged to close by a manufacturer: for example, if it is a low-volume dealership in a declining farm area. The cost of company inventory finance and supervision policies is high, and the manufacturer might prefer superior but less convenient service at a larger and more distant dealership with a higher turnover. As machines become somewhat more sophisticated, with more complex engines and more elaborate hydraulics, there is some advantage in larger and better equipped service shops. Also, holding adequate parts inventory becomes more difficult as larger numbers of increasingly complex machines of different vintages have to be repairable. Further, farm operations have become more critical with respect to timing due to the increasing use of ancillary inputs such as herbicides, pesticides, and fertilizers, and delays due to incompetence or slow parts availability are more costly. Finally, better transport for farmers has decreased the offsetting disadvantage the larger dealerships must suffer by asking customers to travel longer distances.[19]

This position is substantiated by testimony given before the RCFM by several major manufacturers which suggests that in the late 1960s a dealership of minimum efficient size for Canada has been between $170,000 and $200,000 (Can.)

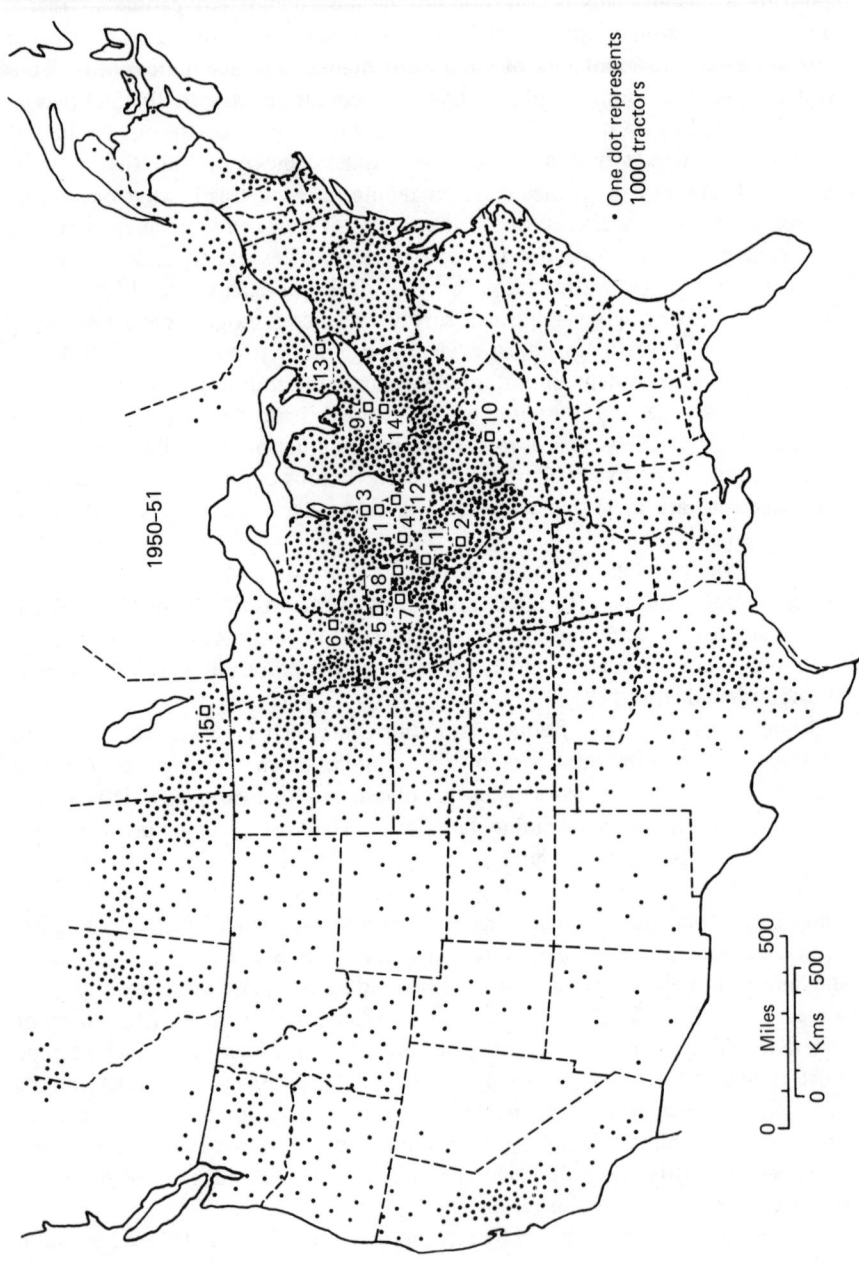

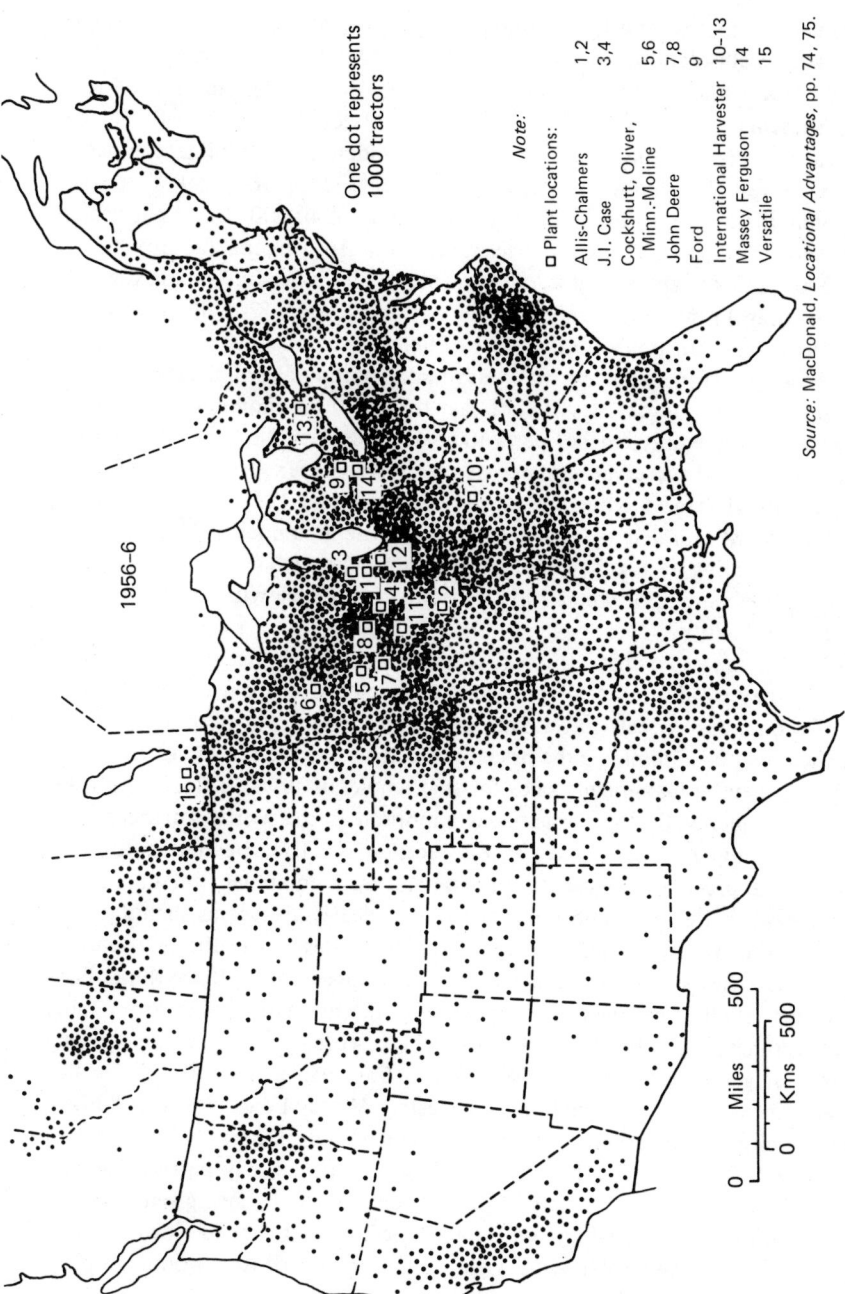

One dot represents
1000 tractors

Note:

□ Plant locations:

Allis-Chalmers	1,2
J.I. Case	3,4
Cockshutt, Oliver, Minn.-Moline	5,6
John Deere	7,8
Ford	9
International Harvester	10–13
Massey Ferguson	14
Versatile	15

1956–6

Source: MacDonald, *Locational Advantages*, pp. 74, 75.

Figure 5-1. The Distribution of Tractors, the United States and Canada

in annual retail sales. The figures appear to take into account a "typical" degree of geographical isolation, and an estimate for the U.S. would probably have been similar. Unfortunately recent complete data on the size distribution of retail establishments are not available for either Canada or the U.S. The RCFM did find the size distribution, however, for five (unidentified) full-line companies in Canada for the period 1966 or 1967. While 22 percent of all dealers accounted for 50 percent of sales (approximately $240,000 (Can.) per dealership) and the next 20 percent accounted for 16 percent (approximately $136,000 (Can.)), the smallest 62 percent of all dealership (approximately $52,000 (Can.)) provided only 30 percent of all sales.[20] Perhaps three-quarters of all dealerships were therefore of uneconomic size.

There is no reason to believe that a substantial group of small dealers does not exist south of the border despite the fact that the uneconomic operations are probably a considerably smaller percentage of all dealers. No exact information is available, but the average number of tractors sold per dealer in the U.S. in 1970 was about a dozen while in Canada it was only nine.[21] In both countries consolidation is retarded by the desire not to lose a competitive advantage to firms consolidating at a somewhat slower pace in cases where distribution costs are not overwhelming or where obviously superior service will not be provided by the larger facility.

White has suggested that "forcing" a limited number of dealers to take more cars than they would independently choose to sell is a means whereby exploitation of economies of scale in retailing can be married to control of dealer profit with the effect of minimizing the price final purchasers face for any given wholesale price, thereby maximizing sales for that wholesale price.[22] There is little qualitative evidence that this practice is important in the North American farm equipment industry or has been in the recent past. The apparently different extent of the practice between automobile and farm equipment retailing might help to explain the differences in variation of dealership versus manufacturer returns in the two industries noted earlier. Nevertheless, from the few years of comparable Census data which are available, it appears that, if anything, farm equipment retailing has been less profitable in recent years than auto retailing. This phenomenon can be reconciled with lack of forcing and even the currying of favor of good dealers by another look at Canadian experience. The major full-line firms (Deere, International Harvester, Massey-Ferguson, and White—earlier Cockshutt) together reduced their franchises in Canada from 2,963 in 1955 to 1,609 in 1969.[23] Again, the U.S. experience may be less dramatic in percentage terms but is likely to have been similar. In any given year the average returns to dealerships were undoubtedly being pulled down by the very large number of dealers who were in the process of discontinuing operations. Nevertheless, manufacturers who acted most decisively to reduce their dealer body found the outlets selected for continuation doing very well. Deere, which led the way in the U.S. to create fewer and stronger dealers, reduced its

U.S. dealer body from 7,000 just after World War II to 3,550 in 1965.[24] In the latter year it reported that a "typical" Deere dealer earned more than 20 percent on equity before taxes.[25]

Exclusivity and Full-Lines

As in the automobile industry (except for company-sponsored dual dealerships, such as Lincoln-Mercury), the franchise is, in effect, an exclusive one in most cases. In 1959, only 8 percent of all U.S. tractor dealers sold more than one make of tractor,[26] and, as the cars, the smaller makers tend to tolerate the arrangement more often than do the larger ones.[27]

Although the practice of exclusivity has been attacked in U.S. courts,[28] the manufacturer has a strong defense in his claim to need a dealer who gives "adequate" attention to his product. In addition, there are numerous pretexts for cancellation and means of harrassment with the effect that exclusivity remains the prevailing practice in good farming areas. On the other hand, even in Canada where exclusivity is legal, joint dealerships are found to be in manufacturers' interests in sparsely farmed areas.

The "tractor dealer" is virtually never just that, but a very high percentage of the additional products he sells are made or at least distributed by the same companies that produce tractors,[29] and, in addition to exclusivity, "full-line forcing" has been attacked in the courts. This is a practice by which pressure is put on the franchisee to handle the complete array of products distributed by the franchisor, and those alone; this may well be to the detriment of other manufacturers and wholesalers and to the profits of the franchisee himself. Nevertheless, although exact figures are not available, the handling of competing products seems to be more common in combines, the second most important product by value, than in tractors.[30] As the importance of the product drops, so, too, apparently does the prevalence of exclusivity.

All North American-based full-line firms, except Ford, produce a broad range of industrial equipment of which slightly modified farm tractors are only a part and have largely kept farm and industrial franchises separate, although selling the complete range of equipment is not unknown.[31] Ford, only arguably a full-line by 1970 (see Chapter Eight), has always allowed its dealers to sell as the market dictates but confined its industrial offerings almost exclusively to modified farm tractors. As is the case with cars and trucks sold jointly with farm equipment, what is an optimal pattern of retail outlets for one line is almost never optimal for the other, and this must be weighed against economies of scale in distribution.

Dealer and Purchaser Credit

In 1961 an article in *Fortune* noted that in the farm equipment industry, "The credit terms both to dealers and to farmers are probably the most liberal in all industry."[32] This situation has deep roots. In one form or another, the manufacturer has always financed the bulk of dealer tractor inventories in North America, and, in addition, retail credit extension through the acceptance of farmers' notes well antedates the introduction of the tractor.[33] The tractor was first sold on a massive scale in the early twenties and was typically delivered on consignment. The retailer had no particular incentive for inventory control, nor did the following up of payments on farmers' notes generally fall into his domain. In an effort to increase dealer responsibility, a system was introduced by most manufacturers in 1924 under which title did actually pass to the individual dealer, but much of the company's role did not really change.[34] It was still necessary to extend liberal credit to the dealer; this would take the form of making the inventory held by dealers interest-free until the farmers' buying season had passed, and a further opportunity was given to carry over at least part of the unsold inventory through the next season at no interest charge.[35]

Although country banks were of some assistance to dealers in the late twenties and again for a few years after the war, loans against farm machinery have never gained great popularity in most rural areas. The companies' acceptance of retail paper in the thirties made the system virtually that of consignment,[36] while in the fifties and sixties an increasing part of retail purchases was financed directly by the companies or by their credit subsidiaries.[37] It is probable that half of all tractors sold in North America were being financed by the companies or their credit subsidiaries in the late sixties.[38]

Before it is sold at retail a dealer may hold a tractor interest-free or "floor-planned" for as long as 14 months, and a combine up to 24 months.[39] Average tractor inventories at all stages of distribution in the late sixties were over 60 percent of annual sales. Of this over half were in the lots and showrooms of dealers.[40] While in practice this only meant perhaps three new tractors at an average dealership, such policies represented a substantial promotional expense. Furthermore, many tractors are specially ordered from the factory with the occasional result that floor-planned machinery must be returned to the manufacturer. A detailed investigation of the expense of wholesale finance including floor-planning will be made in Chapter Twelve.

In evaluating distribution as a barrier to entry into the North American market, one is struck by the formidable linkages between full-line firms and the scarcity of able and experienced dealers. Because the largest manufacturers would usually look with disfavor upon the sale of competitive tractors by their dealers, a newcomer is in a better position to woo good dealers away from their previous lines if he can offer a substitute for the entire range of machinery that is forsaken. To engage in full-line production, however, a firm would have to employ a vastly greater amount of capital than was estimated in Chapter Three. Further, even a foreign firm with a range of implements for use in its home market would find the international differences in agricultural practice and the huge internal variation in farming conditions and crops within the North American market a truly formidable problem. This suggests the alternative of either subcontracting with established North American implement makers to broaden the line beyond tractors or the distribution by manufacturers only to the whole-sale level where a broader line could then be "packaged" by those experienced in selling implements in the North American market.[41]

Another approach for the tractor-maker would be to rely on the dealer to assemble his own full-line, relying on one source for tractors, another for com-bines (the most likely candidate here is the New Holland division of Sperry Rand), and still other "short-line" wholesalers for the myriad of smaller machines and tools offered by a comprehensive dealership. A less attractive alter-native for a dealer would be simply to concentrate on a smaller range of equip-ment and hope that the allure of the new tractor line would make the business more profitable than before.

Posing the issue in this fashion should make it clear that the attraction of a new tractor line alone for most established dealerships would be weak. The problem might be reduced but could scarcely be eliminated by offering the dealer a more complete package of machinery; the market impact of the package would be unknown, and risk aversion by dealers could be expected to be high. The entering firm would thus often be forced to offer its wares through outlets as unknown to potential customers as the machines themselves. The farmer could be expected to buy equipment apparently similar to that offered by familiar dealers and manufacturers under the latter circumstances only at very favorable prices.

DISTRIBUTION CONDITIONS OUTSIDE OF NORTH AMERICA

Tractor distribution outside of North America at the end of World War II was in an inchoate state, with machines sold through a variety of outlets of different types: company stores, automobile dealerships, consignment agents, coopera-tives, and independent implement dealerships. As the years went by, the major firms in each of the markets created distributional arrangements which them-selves had considerable structural importance, and the changes on the whole were continuous and significant. Therefore, only the most minimal discussion of tractor distribution conditions outside of North America will be presented at this point, with a discussion of actual developments reserved for Chapters Seven and Eight.

In North America, manufacturer wholesaling of full-lines through exclusive dealerships seems to be a rather unified phenomenon and was the historical outcome of economies of scale in the manufacture and distribution of farm implements, the necessity for direct contact with dealers on credit matters, and an attempt by strong firms to gain maximum exposure through the best dealers, even before the tractor became an important farm machine. Economies in the use of common plant for some implements are quite important for the full-line

firms, although it was argued in Chapter Three that production complementarities between implement and tractor manufacture are not strong. Distribution economies accruing from selling a full-line for a tractor-maker of any
given size were estimated to be minor, but not insignificant, and were largely a
function of the dispersed character of the market which dictated scale
economies in the operation of regional warehouses, in the handling of inventories stored there, and in dealer supervision. The cost of distribution at all
volumes was found to be quite high, and a large part of this expense was in the
provision of credit, which is subject to considerable scale economies. It was
further argued in the previous section that the continuous depletion of entrepreneurial talent from the sparsely settled areas in which most farm machinery
is sold made the well-functioning dealership the established firm's most important competitive weapon.

Without detailed knowledge of market conduct outside of North America,
could one make confidently-held predictions about foreign distributional developments and their structural significance? First, where firms selling full-lines
in North America entered foreign tractor markets and became important factors
there, one would expect an attempt at full-line selling, regardless of economies
of distribution, simply because these firms already had production and
marketing knowledge for a broad range of implements and would obviously
want to sell at least as much of their product line as did not require substantial
adaptation to local conditions. Their insistence on exclusivity, as in North
America, however, would tend to be a function of their actual strength in the
market at any given time. Where firms not based in farm machinery became
dominant in tractors, however, full-lines might not develop, although these firms
would still favor exclusivity. When firms of both types played important roles,
it would be expected that any gains by the full-line firms would tend to induce
a broadening of the offerings of the other firms.

The argument above assumes that dealerships, once well established, are at
a competitive advantage, whether entrepreneurial or technical talent is in short
supply or not. Clearly, where it is, the competitive advantage to firms selling
through established dealers is much greater. How would one predict dealer
scarcity? One summary measure might be useful for predictive purposes with
respect to dealer scarcity and costs and economies of scale in distribution.
Table 5-2 gives a view of tractor "density" by a crude measure of tractors per
1,000 hectares of arable land. (Obviously, it would be desirable to have a measure which showed the dispersion of the tractor park and its relation to major
population centers, as does Figure 5-1 for North America, but the figures are
suggestive nonetheless.) On the basis of these data one would guess that other
things being equal, the entry problems of costs and economies of scale in distribution *and* entrepreneurial scarcity would be far more serious in North America
and Australia than in any of the other countries in the study.

High tractor "density" is a favorable factor for entry, all other things held

Table 5-2. Tractor "Density"

	Tractors Per 1,000 Hectares of Arable Land			
	1967	*1956*	*1954*	*1950*
North America	24		21	
United Kingdom	62	52		39
France	57	20		7
Germany	154	62		16
Italy	34	11		3
Australia[a]	10		7	

[a]Includes New Zealand.

Source: FAO, *Production Yearbook*, various years.

constant, because both the level of cost and the economies from larger operations are presumed to diminish with geographical compression of the market (for which tractor "density" is a proxy). It is still important to stress, however, as was done in Chapter Three, that both actual distribution cost at any unit sales level and economies of scale are a function of the type of distribution system which become dominant. Geographical compression may allow for certain economies in the handling of wholegoods and inventories at the wholesale level under virtually all circumstances, but economies in the manufacturer-dealer interface depend on how sparsely dealerships can be placed to compete effectively, how much business can be generated from each one, and credit practices.

In North America (and one presumes Australia) the retailing of tractors and other farm equipment usually means permanent accommodation to physical isolation and rural styles of life, and the "take it or leave it" character of full-line exclusive dealing when combined with a shallow pool of established business talent (which we have demonstrated for North America but are only hypothesizing for Australia) raise truly formidable barriers to entry, although full-line exclusive dealing anywhere raises entry barriers to newcomers which could otherwise increase their penetration gradually through established channels.

As late as the mid-fifties, neither France nor Italy were high in density, but, as subsequent growth suggests, those low figures are clearly not an index of extensive farming methods but simply indicate lack of mechanization. The relation of "density" to distributional economies for these countries in earlier years is therefore uncertain, and there may be little, if any, relation to the problems of entrepreneurship or technical support suggested for North America and Australia.

Chapter Six

Overall Conditions of Entry

Before summarizing the apparent entry conditions in various markets, it should be noted that at least three main types of entry can be distinguished. An entirely new, presumably domestic (see p. 14), firm may begin production; a foreign firm may establish a manufacturing subsidiary; or a foreign producer can export into a new market and work toward the establishment of a distribution network. More complicated cases are obviously possible. For example, a foreign firm may buy a share in, or all of, an established domestic competitor, a move which may have very different consequences for the domestic market than a purely domestic change in ownership. Alternatively, an entering firm may source some parts within the new market, either by direct production or by purchase from local suppliers and complement these parts with others imported from the firm's operations elsewhere.

CONDITIONS OF ENTRY INTO THE
NORTH AMERICAN MARKET

In *Barriers to New Competition,* tractors are placed with automobiles, cigarettes, liquor, typewriters, and fountain pens as an industry with "very high entry barriers."[1] All also had very high concentration. Bain's predictions made for this kind of industry were 1) prices substantially in excess of minimal costs with conspicuously high excess profits as a result; 2) substantial selling costs for advertising and/or distributive facilities; and 3) relatively stable industry structure with very little entry occurring over time (unless occasionally through major innovations in product).[2] In a closed economy, all of these predictions are reasonable. In contrast, arguments developed in this study will make it clear why real and potential entry from abroad have contributed to the failure of the first hypothesis to be confirmed in the case of tractors. Viewed from the standpoint of a domestic entrant, product differentiation barriers became no lower subsequent to the early fifties. Large-scale production economies either became greater or perhaps were misestimated in the earlier period, and they must be compared with a much smaller unit market, resulting in a far more formidable economies of scale barrier than was suggested by Bain. Capital requirements were large for production and increased greatly, and trade practice by the late fifties necessitated capital for distribution of approximately the same order of magnitude as for production. Distribution capital was ignored by Bain, perhaps because dealer credit and inventory were at a minimum during the post-war boom.[3]

At the time the research for *Barriers to New Competition* was done, industrial shortages prevailed in much of Europe; this had been preceded by war and before that by a modest role of industrial imports into the U.S.—partly due to U.S. commercial policy and partly due to the paucity of attempts to sell such goods in the United States on the part of foreigners. It was perhaps not seriously misleading to concentrate almost solely on competition from domestic sources for most of Bain's 20 industries. Even in the early fifties, however, low-cost tractor production was taking place in Britain; in the event, as will be seen in the following chapter, the largest part of such activity was being conducted by firms already in the U.S. and Canadian markets. This fact is somewhat fortuitous, however, and immediately raises the question of why the firms involved did so little British sourcing at the time. This puzzling question will be explored in Chapter Eleven. In any event, the ambiguity of the term "minimal costs" is suggested.

The firms just noted, with production on both sides of the Atlantic, may have been the only ones experiencing really low production costs at the very beginning of the period under review, but producers everywhere have never faced tariff barriers to entry into North America, thus allowing full benefit to their foreign production volume. This benefit, however, is partly offset by North American demand for diesel-powered models larger than those sold elsewhere and gasoline rather than diesel power throughout much of the less powerful part of the market in which foreign machines would otherwise be appropriate, thus making some modification necessary.

Transportation costs appear to have been minor throughout and to have become increasingly less important as an impediment to international trade in tractors. Although at any point in time transport costs go up roughly in proportion to tractor size (measured in physical volume), general freight rates in money terms appear to have fallen over the period, while unit values shipped have gone up. In 1961, the declared cost of shipping a 30 h.p. tractor from Britain to the U.S. Midwest was $168 and to the East Coast $145,[4] while in 1968, the charge was only $93 to Montreal.[5] Comparing East Coast freight with the suggested U.S. retail price of similar International Harvester tractors in this size class, and which were actually imported into North America from Britain, the freight charges are found to be 5 percent of the price in the first period and 3 percent in the second. In the later period at least, the rates to Australia from Britain were virtually the same as to North America.[6] It should also be stressed that the figures quoted are general freight rates, and the contract rates actually obtained by an established importer could have been substantially lower.[7]

The product differentiation advantage of established firms remains substantially unaffected when foreign competition is considered. No foreign makers are of sufficient reputation to turn many heads in North America, foreign agriculture in general is unimpressive, and organizing manpower for nationwide distribution would be at least as great a challenge for a foreign entrant as for a new domestic firm. The foreign entrant would be confronted not only with working in an unknown milieu, but in particularly provincial and internally varied parts thereof. Furthermore, there would be the great disadvantage to an entrant offering only a tractor line and, at least initially, possibly an incomplete one at that. A foreign entrant might arrange to complement whatever appropriate implement offerings he had established in other markets with the products of new suppliers, perhaps from within North America, but the organizational problems would be formidable for the establishment of anything close to a full-line of dozens of pieces of equipment for various crops.

Despite these apparently formidable barriers, it is nevertheless quite possible that international competitive factors are in part responsible for the apparent failure of Bain's predictions concerning price-cost margins and profitability. The factual material on these issues is reserved for Chapter Twelve, but some of the relevant theory can be outlined here. Figure 6-1 is an attempt to schematize some of the entry conditions of the North American market; it is based on a diagram used by Bain to illustrate product differentiation barriers in the presence of scale economies.[8] The position of dd, the demand curve of a potential entrant, is a function of the price set along DD, the demand curve of an established industry participant when established firms change their prices together. In the situation shown $d_e d_e$ is the (probably quite inelastic) initial demand function facing the entrant at the established firms' maximum entry-deterring price, P_L. The dashed line ac_p represents production costs for a North American producer, while his total costs employing the prevailing distributional practices is ac. If a newcomer had no other relevant production and could enjoy production costs at any volume similar to those of established firms (and capital cost considerations of the ordinary kind could make them higher), then the price-cost margins for established firms would be extremely high.[9] It is probable that this adequately represents the situation facing a new producer located within North America contemplating offering a range of wheeled tractors.

The international economy offers two obvious but possibly essential emendations to the picture. Costs curves ac_p' and ac' are drawn on the assumption that a potential entrant from abroad is already experiencing sales there of

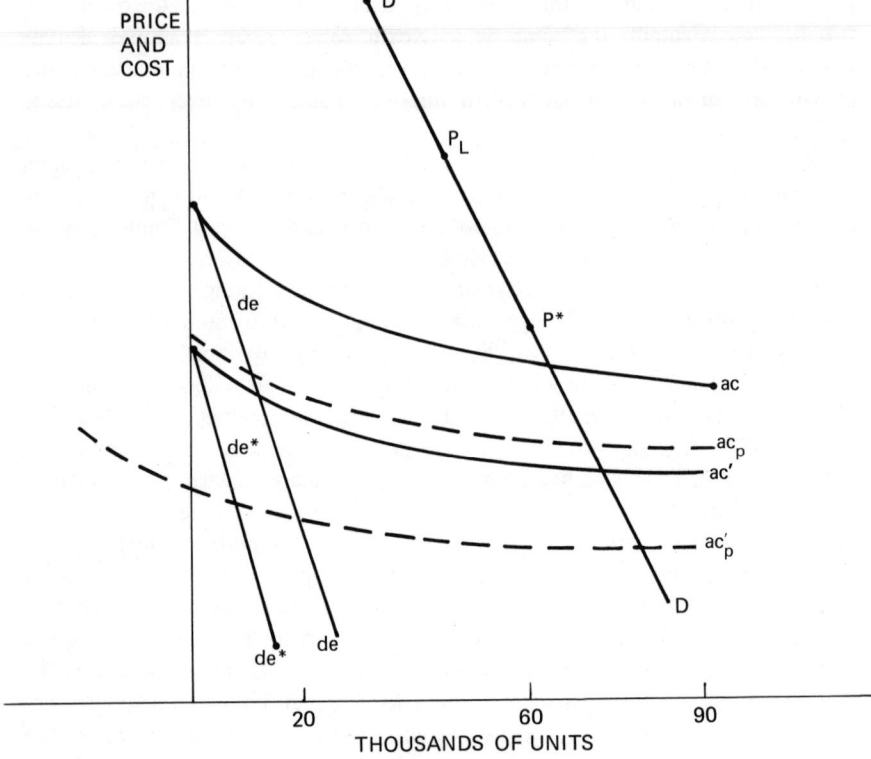

Figure 6-1. North American Entry Conditions

20,000 and at an average cost of production of approximately 75 percent of that facing a North American firm at any volume. It is important to note that the foreign entrant has no distributional cost advantages in the new market, however, and ac' is drawn on the assumption that his absolute amount of distributional cost at any volume is similar to that of North American based firms (in fact, there would be less capital involved in the lower value of inventory but this would be set against a minor cost disadvantage of not being a full-line firm and possibly also a higher cost per unit of capital). It should be stressed that although the estimated economies of scale in the tractor industry are approximately reflected in the diagram along with the relative magnitudes of distribution and production cost for the range between 20,000 and 90,000 units, the cost curves up to 20,000 units are nothing but illustrative free-hand sketches, because no estimates are available. In the situation as presented, only at price P^* and its associated entrant demand function $d^*_e d^*_e$ will foreign entry be deterred. The story is simplified by assuming that both a foreign and a domestic newcomer would greet the same volume of trade at similar prices. Either antipathy towards foreign supply or positive marketing spillover could make a difference, of course, and the former may be more likely for at least some potential entrants into the North American market. Difficulty of entry for the foreign firm is greatly determined by the shape of the distribution cost function in the range under 20,000 units where actual estimates are lacking (and, more generally, by the relative importance of production and distribution costs).

This broader look at the conditions of entry has obvious implications for economies of scale as a barrier to entry as well. The entire notion of that barrier concerns the relationship between increasing market share and falling production costs. The entrant is assumed to face unit costs at any given volume no lower than established firms and to depend for his total sales only on the market in question. The barrier necessarily implies a non-negligible scale of entry and therefore makes necessary some conjectures about the price and output reactions of established firms. An international entrant in the circumstances sketched above skirts this problem. Initial entry may well be possible on a very small scale despite the overwhelming economies of scale barrier as defined by Bain if the foreign supplier has either lower costs at any volume or has relevant production elsewhere. In the latter case it is obviously possible for an entrant to have even higher unit costs than an established firm at a given volume if the total scale of operations of the newcomer is great enough.[10] Another consideration could be of importance: uncertainty about the movement downwards or leftwards of ac'_p might retard investment in new plant and machinery by established firms and further reduce achieved costs by comparison with what was technically possible over time.

The general point here appears to be an important if intuitive one: advantages from foreign activity, particularly low production costs from whatever source, can theoretically negate any other disadvantages which might otherwise serve to bar entry. If entry were to be observed into the North American market, it would presumably be either from the largest firms elsewhere or ones located in countries with the lowest absolute costs. A strategy which could complement such a move would be the purchase of an established North American tractor-maker—a move which could both overcome product line and distribution problems and establish a "domestic" manufacturing image, whatever the actual amount of North American value-added. Any possible trade-off between domestic content and market penetration could easily be incorporated into the model without changing its essentials.

CONDITIONS OF ENTRY ELSEWHERE

If economies of scale are as important in all of the markets as they are believed to be in North America and the U.K., then one would posit very high entry barriers for any newcomer relying wholly on new domestic production, whether

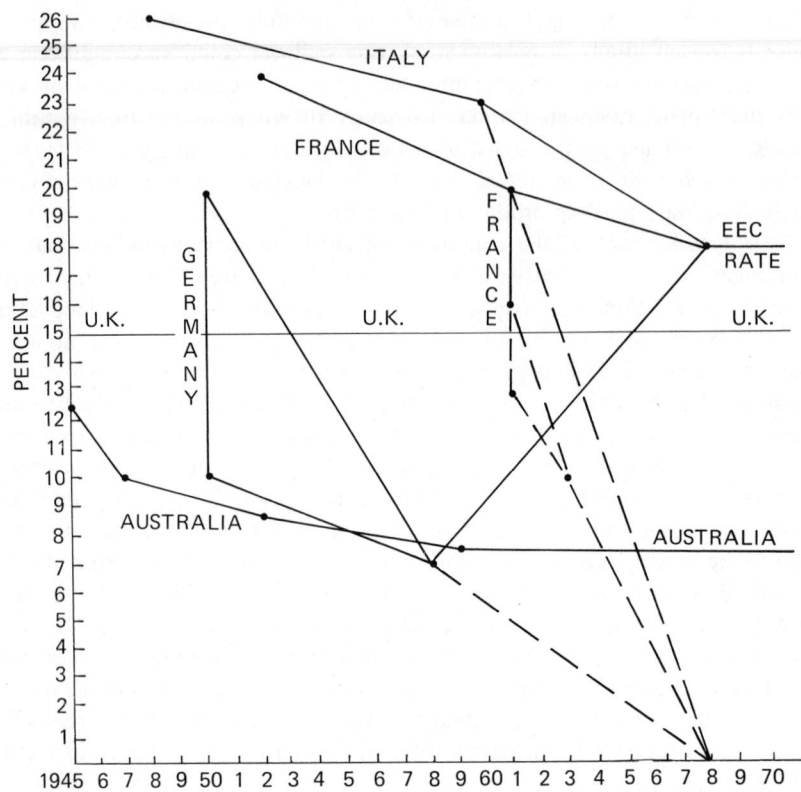

Figure 6-2. Nominal Tariff Rates for Wheeled Tractors

[a]The Australian preferential tariff which applied to British imports throughout was zero.

Notes: Vertical distances at any one point in time indicate differing rates according to tractor specifications.

Dotted lines indicate intra-EEC rates.

U.S. and Canada had no tariff on tractors for farm use. The U.S. had an 11.5 percent rate on industrial use tractors.

Sources: The basic source for nominal tariffs in the late fifties is P.E.P., *Atlantic Tariffs and Trade* (London: Political and Economic Planning, 1962). In addition, the following material has been used on tariff levels for some countries. France: Linneman, "The U.S. Tractor Industry," pp. 46–49, 56–58. Germany: *Farm Mechanization,* April, 1951, p. 158; *FIMR,* October 1, 1955, p. 1083; May 1, 1959, p. 1830. Italy: Neufeld, *Global Corporation,* pp. 353–54; *FIMR,* June 1, 1967, p. 558. Australia: Linneman, "U.S. Tractor Industry," p. 119; *Farm Engineering Industry,* September, 1969, p. 333.

domestic or foreign-owned, once any firm or firms attained a substantial volume of sales. The latter would be presumed to develop during the initial period of rapidly expanding domestic demand which was outlined for the various markets in Chapter Two. How does consideration of foreign production amend the picture? In most of the discussion of North America, it was simply assumed that foreign entry, if any, would be the marketing of imported whole machines (perhaps "knocked down" for transport economies) because of the importance of economies of scale, the absence of tariff barriers against machines for farm use, and modest transportation costs. The situation elsewhere has been complicated by trade controls during the early post-war years and by tariff barriers throughout. Imports from the dollar area were severely restricted in all of the non-North American markets in the study until the late fifties. All of them necessarily saw their "tractorization" realized from non-American sources of supply.

In addition to restrictions on dollar trade, France had considerable tariff protection and directly controlled trade in tractors until 1956.[11] Most other markets also have had considerable tariff protection which has blunted the ability of any firm to achieve scale economies by selling in multiple markets. The varying rates are presented in Figure 6–2, and comments about the levels of effective protection afforded are appended thereto.

What other relevant information do we have about non-North American entry conditions for foreign production? Australia was estimated in Chapter Three to be a high-cost country and has had only modest tariff protection (despite the government's determination to have some heavily subsidized domestic produc-

---

Effective rates of protection of some relevance have been estimated by Bela Balassa, "Tariff Protection in Industrial Countries: An Evaluation," *Journal of Political Economy*, December, 1965, pp. 573–94. Balassa estimates the effective rates only for agricultural machinery as a whole, but for Britain, the overall nominal rate of 15.4 percent is very close to the nominal rate for tractors which comprised 70 percent of actual U.K. production at the time the study was done. His estimate of the 1962 effective rate for farm machinery is 21.3 percent. No estimates of effective protection for individual Continental countries were done, but Balassa reckons the final EEC overall farm machinery rate of 13.4 percent corresponds to a 19.6 percent effective rate. An indication that the EEC effective protection rate for tractors may be proportionately yet higher is indicated by the case of the technologically similar automobile industry. Here a 36.8 percent effective rate corresponds to a 19.5 percent nominal rate. For the United States, the virtually nil farm machinery nominal rate corresponds to an effective rate of *minus* nearly seven percent. Australian effective protection cannot be even roughly estimated, because the net effect of the subsidies described in Chapter Three and Australian components produced under heavy protection is not known.

tion). On the other hand, it was hypothesized that Australia has had the same kind of product differentiation distribution barrier as has North America and probably to a greater degree, and the expense of small-scale distribution could be even more severe. It must be remembered that the entire Australian market has usually been below the size at which economies of scale in distribution have even begun to be estimated for North America. In addition, concentration has been high throughout, and one might therefore predict as little or less successful entry than in North America after the first few post-war years, especially for non-British firms which would face the 7 to 9 percent tariff.

For Britain, low estimated internal cost levels, in addition to protection, would appear to have made it a very unlikely candidate for entry from foreign production, at least from the other markets in the study.

Less can be said about the remaining three markets: Italy, France, and Germany. Linneman has established that, by the time convertability was restored, the cost advantage of some European, particularly British, production over North American production was freely admitted by U.S. executives,[12] and so no entry sourced from U.S. plants would have been expected during any part of the post-war period. In Italy, in addition to the very high level of tariff protection, the Fanfani finance plan would have put foreign manufacturers at a very great disadvantage unless they were prepared to provide retail financing themselves because rural credit not under government auspices was exiguous. What one would expect would be an increase in firm interpenetration among the three Continental countries (and a rationalization of production among them by any one firm) with the development of the EEC. This seems particularly likely because the product differentiation advantages to established firms, at least as regards distribution, should be much less significant on the Continent than in North America and Australia as should be diseconomies of small-scale distribution.

Chapter Seven

Entry

This chapter will discuss actual entry and subsequent strategy in the world trac-tor industry. Although we will begin to discover the reasons for the success or failure of the various entering firms and their competitors, only after a complete discussion of firm behavior, to be further explored in the four chapters that follow, will a complete picture of differing competitive strengths emerge.

Chapter Eight will discuss aspects of overall competitive behavior (other than product design) and the impact on industry structure of that competition. Chapter Nine will present a detailed examination of product behavior, narrowly construed as the physical attributes of the machines sold. Chapter Ten will discuss issues related to international differences in price structures and levels, and Chapter Eleven will sum up the apparent overall strategies and strengths of the various firms.

The preceding chapter concluded with a discussion of entry conditions by market. Before examining the entry which did in fact take place, it is appropriate to offer a few additional a priori hypotheses about the world industry as a whole. First, one would predict that at least some North American-based firms with sunk costs in the attainment of marketing and production knowledge,[1] would be tempted to begin tractor manufacture in other markets. The North American firms would have yet another advantage in the marketing and produc-tion experience with implements to complement tractor sales. Indeed, some North American firms were well established in foreign implement markets before the war. Somewhere within the North American land mass there had developed demand for implements broadly suited to virtually all types of agriculture, and North American firms could be expected to sell as many of these implements along with their tractors as possible. Experience with other industries suggests that those firms most likely to enter into foreign production when exporting became impossible would be those with the largest established stake (in either tractor or implement sales).[2]

Second, given the economies of scale in the industry, production should be observed to be centralized as much as other factors allow. The latter would include: trade barriers or prohibitions, the necessity of continually adapting the product to local conditions (a factor which the discussion in Chapter Four would discount and which will be further diminished in Chapter Nine), and a significant marketing impact attaching to production near the point of use, a consideration alleged to be particularly important for producer durables.[3] Given the conservative nature of agriculture in general, this latter point is one about which a priori prediction is difficult, and the importance of which remains to be established.

Third, the foreign purchase of an existing firm in an established market might sometimes be observed, either to gain production facilities or to overcome product differentiation disadvantages. This form of entry would clearly be more attractive where established distribution is important or the marketing strength of the acquired firm is strong for some other reason, where the centralization of production facilities is made difficult by tariff barriers, or where local production per se is thought to contribute strongly to marketing impact.

Fourth, large firms should have an enormous advantage over smaller ones such that firms of modest total volume would be observed operating outside their own home markets only if the manufacturing base were an extremely low absolute cost area. Indeed, one would predict difficult times for small firms versus large ones with similar factor costs everywhere.

Finally, as already suggested, exclusivity would always be desired, and presumably won, by the most important firms in each market, and where North American firms became important competitors, one might expect defensive product-line broadening and imitation of other marketing techniques by their adversaries.

Because there was some production, however modest, in all markets prior to the war, all new post-war participants can in some sense be regarded as entrants into existing markets. Nevertheless, as Chapter Two made clear, immediate post-war entry into any market outside of North America was tantamount to getting in on the "ground floor" of the market's development.

NORTH AMERICA

Of the seven major firms selling tractors in North America in 1937, all of them offering a full-line, every one was still in the market in 1970 (although Oliver and Minneapolis-Moline came under common ownership in the early sixties). Furthermore, only the Ford-Ferguson entry in 1939 resulted in a substantial share for a newcoming firm. Exact sales figures are not usually revealed by firms participating in the North American market, and Figure 7-1 presents nothing more than official data for U.S. shares in 1937, probably quite accurate North American estimates for 1966, and a few scattered estimates for some firms' shares in the intervening period (Canadian sales are under 20 percent of those in the U.S.). The reasons for the varying shares of the firms are many, but as was predicted, they are not largely attributable to sustained price difference. During the sixties, for example, the major firms were loath to allow their prices to vary from similar models of another brand by 5 percent, and one company suggested that the true figure was only half of that.[4]

The Ford-Ferguson Venture

The Ford-Ferguson entry took place before the war, but its repercussions on machine design and market structure came during and after the conflict; the newcomer captured about a quarter of the market by 1943.[5] The original arrangement between Ford and Ferguson was not in writing, but Ford was to finance and produce the tractor for which Ferguson had distributional responsibility. Unlike the other major companies, Ferguson worked through independent distributors who built up their own dealer system—a nationwide network of 33 distributors with an average of 100 dealers apiece was established in the United States during the war. Henry Ford's decision to let material produced in his factories be marketed by a company over which he had no control was mainly based on Ferguson's long experience with, and devotion to, proselytizing the world for the cause of the "Ferguson System" and training others to do likewise. This arrangement may have seemed satisfactory to the elder Ford, but his grandson, whose job it was to attempt to untangle the company's poorly recorded and failing affairs in 1945, was not anxious to continue cooperation with Ferguson without majority control of the marketing company.[6]

There were many issues involved in the ensuing breach and the subsequent litigation, but the essential points are as follows.[7] An impasse was reached between the two parties in 1947. Ferguson failed to make credible plans for U.S. production of tractors, and this led distributors to abandon him in droves for the newly formed Dearborn Motors which was to market the forthcoming Ford tractor. After a year of confusion, Ferguson established a modest source of supply by using major components from Continental and Borg-Warner, but Ferguson suspected Ford of intervening to make supply and distribution arrangements difficult to attain. Litigation vindicated Ferguson's claims of patent infringement, although the mandatory redesign of the Ford tractor's hydraulic system and implement linkage was not difficult. Finally, Ferguson's allegation that Ford had purposely hindered his independent establishment was not sustained, in part because the years 1951 and 1952, when the suit was finally settled, were good ones for the nascent Ferguson operation.

The post-war boom ended in 1953, and both Ferguson and the Canadian firm, Massey-Harris, found their North American market appeal very weak. Massey-Harris was a company strong in harvesting machinery but undistinguished as a tractor-maker. By contrast, Ferguson, like Ford at the time, was selling only one tractor model and a modest range of implements for use with it. Ferguson

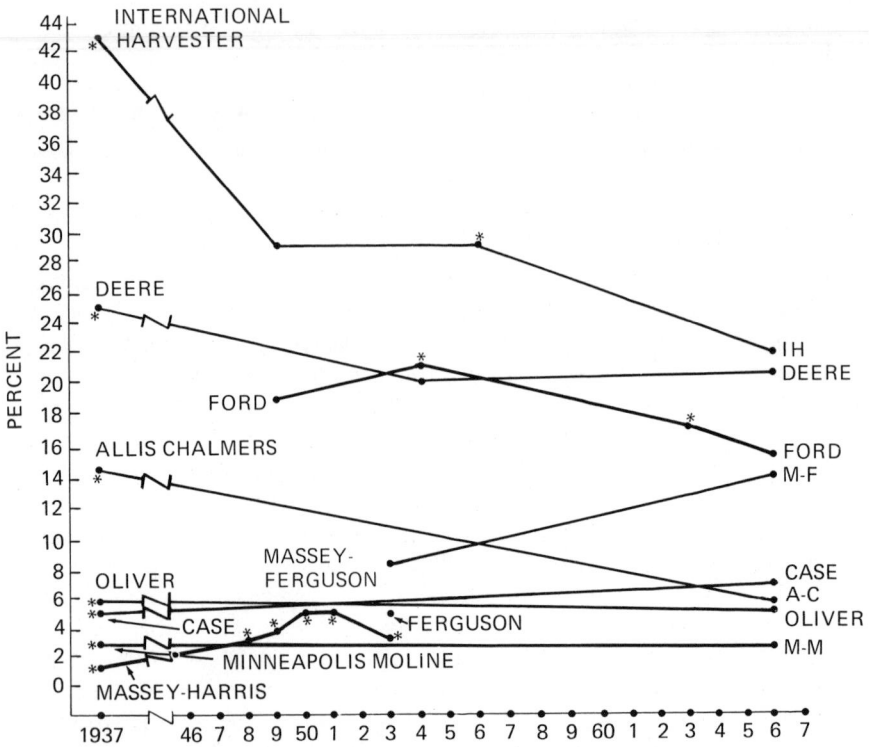

Figure 7-1.   North American Market Shares

*U.S. Market Share Only.

Source: For Oliver, Case, Minneapolis-Moline, and Allis-Chalmers, only two data points are available, those of 1937 and 1966. The former are taken from U.S. Temporary National Economic Committee, *Investigation of Concentration of Economic Power,* Monograph No. 36 (prepared by the Federal Trade Commission) (Washington: U.S. Government Printing Office, 1940), p. 210. The latter are the market shares calculated for Table 1-2. In addition to these data we have the following: Ford and International Harvester estimates for 1949 from Michael Conant, "Competition in the Farm Machinery Industry," *Journal of Business,* 26 (January, 1953), p. 36, which should be taken with caution because they are shares of total U.S. *production* during a period of unusually high exports to areas outside of North America. Neufeld's *Global Corporation* (pp. 61, 66, 121, 139) gives a complete accounting of Massey-Harris' U.S. market share, and sufficient production material for the approximate North American share of Massey-Harris and Ferguson at the time of the merger to be estimated. Ford claimed second-place in 1954 (see Alan Nevins and Frank Ernest Hill, *Ford: Decline and Rebirth, 1933-1962* (New York: Charles Scribner's Sons, 1963),

and Massey-Harris merged in late 1953. In the following years, Ford expanded its offerings of tractors and implements.[8]

Entry During the Post-War Boom

As was the case with automobiles, there was a great shortage of tractors in the first few years after the war, resulting from bottlenecks in the supply of essential materials and parts, labor unrest, and enormous home and foreign demand. In 1947 tractors sold for two or three times their list price.[9] Entrants rushed into both industries, although their establishment in tractors was much easier to accomplish because of the relative simplicity of tractor design. Any "scrounger" who could get an industrial gasoline engine, some transmission parts, and some sheet steel and iron castings could go into the tractor business. It is not known exactly how many firms entered the fray, but at least 22 new makes were important enough to be noted by the acknowledged authority, R.B. Gray, for the period up to 1950.[10] A few of the newcomers claimed to have planned to enter before the war, but most offerings were the quickly marketed sidelines of established firms, many of which weren't based in farm machinery. Some expired as early as 1950, and, although a few lingered on into the late fifties, none ever captured a significant market share. A simple, never better than conventional, and sometimes quite crude, design, produced at low volume and distributed through secondary channels, was severely disadvantaged in competing with the offerings of established producers which by 1953 were well able to meet demand and were about to face a long period of excess capacity.

The only post-war North American-based entrant with any staying power in the conventional tractor race was the Canadian firm, Cockshutt, which added a tractor and a combine to an already impressive implement line and became a full-line producer in 1947. Another minor entrant was the National Farm Machinery Co-operative which produced a few thousand tractors between 1939 and 1952. Its role and fate are best discussed as part of the general activities of cooperatives in Chapter Eight, but part of Cockshutt's penetration of the U.S. market was accomplished by supplying machines to be sold under the "Co-op" label to complement the output of the Co-op itself.[11] Although Cockshutt tractors were not generally regarded as inferior in the United States, neither did they have any unique appeal,[12] and, even after taking over the Co-op's Bell-

---

p. 383) with a 21 percent market share and must have been very closely followed by Deere—this is reflected in the figure. In 1956 International Harvester claimed between 28 and 30 percent of the U.S. market (A.D.H. Kaplan, Joel B. Dirlam, and Robert Lanzillotti, *Pricing in Big Business* (Washington: The Brookings Institution, 1958), p. 70) and leadership in the market. In 1963 Ford claimed 17.3 percent of U.S. unit sales (*Business Week*, October 24, 1964, p. 174).

vue, Ohio tractor plant in 1952, Cockshutt continued to sell most of its output in Canada. The small firm was hit hard by the lean years after 1953, and made losses in four of the five years, 1954 through 1958. Nevertheless, it had tractors of adequate quality and an outstanding combine by the late fifties. The latter is one of the reasons why White Motor acquired the firm in 1962, after which White's Oliver tractors were sold in all of Canada except Québec as Cockshutts. (Oliver was acquired by White one year before the purchase of Cockshutt, and Minneapolis-Moline one year afterwards.)[13]

The Entry of Specialist Firms

Another source of entry, beginning in the fifties, was from small domestic firms aiming their products at segments of the farm tractor market not served satisfactorily by existing machine designs. The huge, diesel, four-wheel drive vehicles which the major companies began marketing extensively in the late sixties were in earlier years sold by several smaller firms which in some cases continued to make them. A successful entrant of this type was FWD Wagner which in 1954 adapted logging and construction equipment for farm work, principally for use in wheat-growing areas.[14] Considering that the annual sales of firms such as Wagner ranged from a few hundred to a few thousand, one might wonder how they existed at all in competition with the major firms. One explanation is that the value-added in their production was extremely low, with virtually no in-house manufacture of major components. The other reason is that price-cost margins for high horsepower tractors, including the four-wheel drive units, were set at very high levels by the major firms, a practice which will be explored in Chapters Eight and Ten.

A Canadian entrant which paralleled the "specialist" newcomers throughout the fifties and sixties in the U.S. was Versatile of Winnipeg. Starting in 1963 with a range of simple implements, it added swathers, then combines, and in 1966, large four-wheel drive tractors designed for the grain belt.[15] Unlike many small U.S. firms producing similar machines, Versatile's experience was with farmers and farm equipment, and it soon became the largest of the "specialist" firms. In 1970 it was reported that output was 12 a day, or, assuming a constant rate, over 3,000 a year.[16] Its location was in a low wage area close to its market which allowed direct shipment to dealers, and it sold through dealerships handling other lines, with higher than usual retail margins. Although Versatile's principal selling point up to the early seventies was price, a combination of price reductions after 1967 by the only two firms which were then in competition with Versatile's models, the addition of several new competitors in the class, and an enormous price rise by Versatile, due to some unknown combination of increased costs (in part due to the introduction of floor-planning) and a more conservative estimate of demand elasticity, made the prices of Versatile machines much less dramatic in 1971 than they were in 1967 (see Table 7-1).

Table 7-1. The Decline in Versatile's Price Appeal (Prices in U.S. Dollars)

	1967 Selling Season			
	Suggested Retail Price	*Drawbar H.P.*	*SRP Relative to Versatile*	*Price/Drawbar H.P. Relative to Versatile*
Case 1200	$17,639	107	191	208
IH 4100	16,774	116	181	186
Versatile				
(118 h.p.)	9,246	118	100	100

	1971 Selling Season			
	Suggested Retail Price	*Drawbar H.P.*	*SRP Relative to Versatile 145/118*	*Price/Drawbar H.P. Relative to Versatile 145/118*
Case 1470[a]	$15,680	132	99/113	109/100
IH 4156	17,540	118	111/127	137/126
Versatile 145	15,795	145	100/114	100/93
Versatile 118	13,841	118	88/100	108/100

[a]Described as a beefed-up version of the 1200.
Source: Barber, *Prices*, p. 58. N.F.P.E.D.A., *Official Guides*, 1967, 1971.

Entry from Abroad

Several Continental, especially German, firms advertised in the late fifties and early sixties, but little was ever heard of most of them after their initial marketing announcements.[17] A foreign newcomer with somewhat greater staying power was Porsche, named for the engineer who conceived the Volkswagen, the leading German export to the United States. Some advertising used Porsche's picture and reference to the Volkswagen and his sports car. The Porsche tractor was of advanced design and offered a lightweight air-cooled diesel engine. It was introduced to the U.S. while it was enjoying enormous success at home, and came in four models in the small and medium horsepower classes. It was complemented by no other equipment, however, and was sold through dealers holding other tractor franchises. Furthermore, advertised prices were not impressive, although dealer discounts may have been somewhat higher than for other machines,[18] thus possibly allowing a somewhat lower final transaction price than could otherwise have been quoted. The Porsche effort, which was based upon a wholly-owned distributor, began in 1958 but failed to achieve any real acceptance in North America. When its market share at home began to decline in the early sixties, complete withdrawal followed, and in 1963 Porsche was sold to Renault (see p. 98).

In 1966, the leading German tractor maker, Deutz, began wholesaling in the southeastern United States, an attractive target for foreign entrants, because it combines rather high tractor density with modest average horsepower by comparison with the Midwest. Distribution throughout the United States and Canada was soon established, and by 1970, Deutz offered five models from 32 to 96 horsepower,[19] thus coming closer than any other import to covering the complete range of American tractor needs, and it also offered dealer financing. There was nothing to indicate that Deutz was doing well, however. As of 1970, it was not a full-line company,[20] so presumably few of its dealers were exclusive, and its broad power range could scarcely have been greeted with enthusiasm by the full-line firms expected to share dealerships. Even more importantly, Deutz tractors were not low priced. Advertising stressed low cost of operation and the virtues of its air-cooled diesel engines, but not price. There might be some attraction in better fuel economy, but it is a feature common to all diesel-powered units, and the additional saving on such items as radiators and water hoses was unlikely to strike buyers as an important extra. In Western Canada Deutz sold its machines through cooperative channels—a practice to be evaluated in the following chapter.

Another foreign entrant which enjoyed only modest success was British Leyland. By contrast with the Continental entrants, British Leyland found at least two excellent established distributors. One of them had been a Cockshutt regional distributor until the firm's purchase by White in 1963, and the other, acquired in 1968, had actually produced a tractor briefly during the post-war boom. Both distributors also handled an impressive array of implements.[21] Press reports indicate that the more recently acquired wholesaler, which became the largest of British Leyland's five U.S. distributors, purchased 1,000 tractors between August 1970 and April 1971.[22] Many British Leyland dealers held full-line franchises as well.

The most successful foreign entrant was the other major British-owned firm, David Brown. Although the company made a stab at introducing its products into the United States in 1957,[23] it was apparently as unsuccessful as the other European hopefuls of the same period. Brown's first major experience with the U.S. market was in 1960 when it manufactured a slightly modified version of its small "850 Implematic" tractor to be sold as the Oliver "500." [24] According to company executives, David Brown was anxiously awaiting an opportune moment to make a more substantial move into the market when Ford dropped its 24 independent distributors in 1964 (see following chapter). While a few of these distributors became Ford branches, 14 of them banded together to market David Brown tractors in addition to a number of implement lines which the firms continued to distribute.[25] As was the case with British Leyland, dealer financing was offered, and wholesale prices on the three small and medium-sized tractors in both lines were from a few percent to over 15 percent lower than the comparable models of leading firms. Yearly estimates of the number of David Brown tractors sold are not available, but an average of over 3,000 were sold in the years between 1965 and 1970, or an average of less than 2 percent of the market. Industry reports, however, indicate that market share grew in each successive year.[26]

A final source of entry was Japan. By the late sixties many U.S. dealers sought to supplement their main tractor lines with equipment from Japan in the under-30 h.p. category, because the major makers had virtually abandoned it (except for lawn and garden equipment). The first firm to recognize this need was Kubota, Japan's largest farm equipment maker, which began extensive penetration of the North American market in 1969, and in 1970 Mitsubishi began North American tractor sales; both firms operated wholly owned distributorships. The National Equipment Dealers' Association, the David Brown distributor system, began selling the Japanese Satoh tractor in 1970. During that year Kubota sold 600 of its products, Satoh 55, and Mitsubishi 150. The closest competition from a U.S. maker was the International Harvester 140 at 23 h.p. which claimed to sell "several thousand" annually at several hundred dollars more per unit. This premium was up to 20 percent of the International Harvester suggested retail price.[27]

In 1970, only 6 percent of U.S. unit sales, or about 10,000 machines, were in the under-30 h.p. category, but the abandonment of this part of the market by the major firms was a rather strong admission that they could not compete, even though there were fewer than 10,000 tractors produced in all of Japan as late as 1965.[28]

The main point to be emphasized about U.S. and Canadian entry is that the impact of none of the entrants since Ford-Ferguson was really substantial if measured by market share achieved. Except for Versatile, the output of the domestic newcomers has probably been in the hundreds, and no importer up to 1970 probably ever sold 5,000 machines in the United States in any one year. Whatever modest success has been achieved seems to have been won by low prices and either a previously established distribution system (David Brown and British Leyland) or an unusual special-purpose design (Versatile). A determined effort to develop distribution without any of these attractions (Deutz) appeared to be a dubious enterprise.

Did the Europeans suffer from importation per se or from the firm or country of origin? Although there is a school of thought which sees the U.S. farmer as resistant to foreign machinery,[29] there is really no evidence to back it up. While he might well be chary of foreign machinery from countries (or firms) with an unfavorable image, this is a distinct issue. One of Ford's most popular models in the fifties and early sixties was the "Major," the British manufacture of which was probably known by every purchaser. The U.K. probably had a "neutral" image as regards agriculture and an acceptable one in manufacturing, and, although the Ford name was an obvious marketing asset, there is little reason to believe that either David Brown or British Leyland suffered any important disadvantage simply because they were imported or imported from the U.K. In the former case, at least, any "spillover" of the firm's home market reputation would have been very favorable because it had a superb reputation for precision manufacture and advanced design. The Deutz case is more problematic; although it advertised itself as the "world's largest producer of air-cooled diesel engines" and Germany's manufacturing reputation was probably considerably higher than Britain's, agricultural productivity in Germany was notoriously low. The net impact was unlikely to have been important in achieving market acceptance.

THE UNITED KINGDOM

Ford and Massey-Ferguson (before 1954, primarily the organization operated by Harry Ferguson) dominated the British market during the entire period, usually selling similar machines at similar prices. Ford was the major British producer before the war, producing 18,000 units in 1937, and it accounted for 120,000 of the 128,000 British tractors produced between 1940 and 1944. Ford's near monopoly during the war was government sanctioned because most other potential competitors were required to devote their full efforts to armaments production.[30]

Immediate Post-War Entry
As late as 1946, before U.K. exports boomed and while two-thirds of all units produced remained in Britain, Ford was still producing 87 percent of the total. Nevertheless, in 1945 Harry Ferguson, whose association with Ford was limited to North America, entered into an agreement with the automobile firm, Standard Motors, to produce a tractor to his specifications. In addition, David Brown, with which Ferguson had a short-lived production arrangement in the thirties and a firm with great general machinery experience, began substantial tractor production in 1947, as did the Nuffield organization (later a part of British Leyland) in 1948.[31] Figure 7-2 shows scattered unofficial estimates of post-war U.K. firm shares.

Although over one-quarter of Britain's need for tractors was supplied from the United States during the war,[32] the domestic industry's capacity and controls brought imports to a virtual halt immediately thereafter. Producers which had formerly exported from North America found it necessary either to write off the British and other non-dollar markets or to establish production facilities outside the dollar area. The expected devaluation of the pound and Imperial Preference made the U.K. an attractive production location. Massey-Harris, International Harvester, Minneapolis-Moline, and Allis-Chalmers, all entered into British tractor production by 1949 in some fashion, but Minneapolis-Moline seems to have abandoned the fight almost immediately.[33]

In addition to this group, a number of small domestic firms produced standard agricultural wheeled tractors in the early post-war years, but they failed to

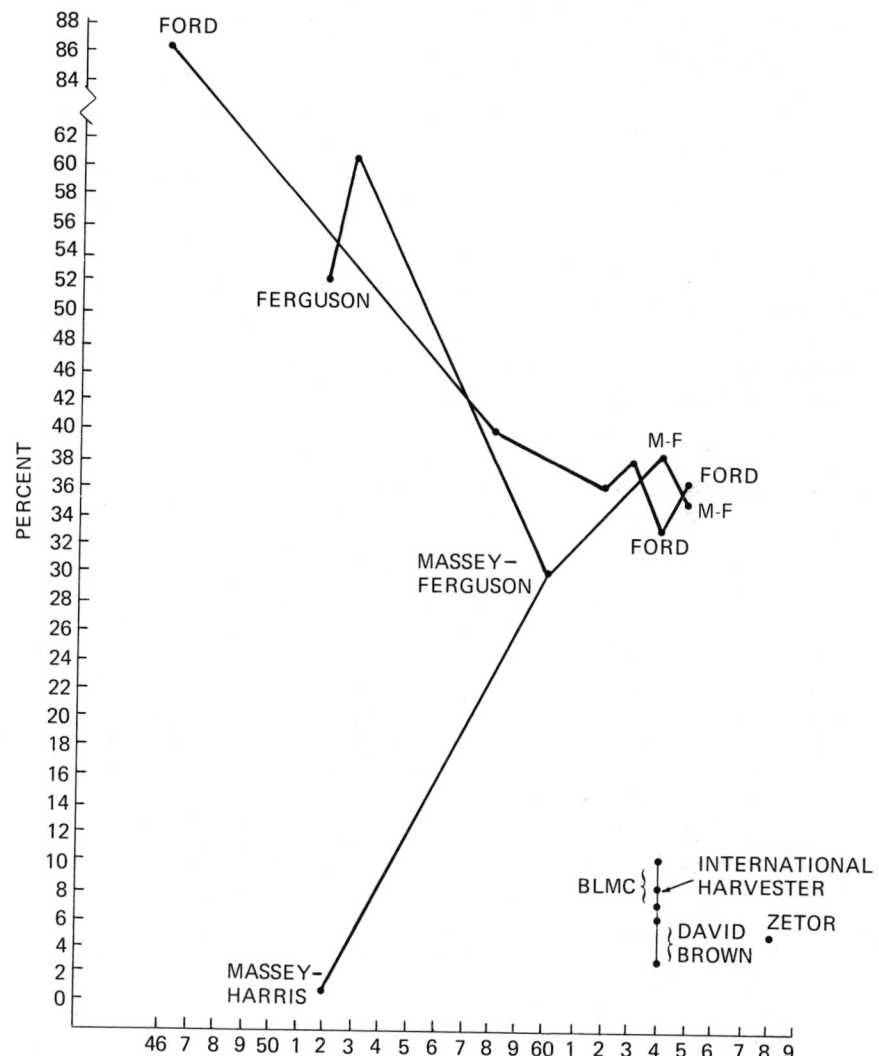

Figure 7-2. U.K. Market Shares

Source: Ford's 1947 share is given in U.S., *European Tractor Industry,* p. 26; data for 1952 and 1953 are from Neufeld, *Global Corporation,* p. 139. Later shares calculated from total sales figures and material given in *FIMR,* February 1, 1959, p. 1480; January 1, 1964, p. 71; January 1, 1967, p. 44; Donaldson, et al., *European Equipment Industry,* p. 10; *AMJ,* April, 1969, p. 68.

establish a non-negligible share in the market for reasons similar to those applying to their North American counterparts.

Attempted Entry in the Fifties

John Deere, in its first recorded attempt to do more than export from the U.S., laid plans to open a factory in Britain in 1952; ground was broken for a plant in Scotland, but the project was abandoned after government delays in authorizing steel purchases.[34] How the delays could have been other than a pretext for a changed decision is unclear.[35] Aggregate British production, over 80 percent of which was exported, peaked in 1951 and then receded for a couple of years; Deere executives may have feared for the long-run viability of the project partly for this reason, although output in the United States was then permanently past its peak.

Deere made another attempt to enter the market from its German subsidiary (Lanz, acquired in 1956) in 1960, but it had negligible market impact.[36] The first sustained, if modest, Deere attack was the offering of its larger North American tractors in 1962, a move taken in conjunction with setting up its own British distribution and jettisoning its former exclusive importer. While the Lanz machines were both conventional and probably overpriced (they were not sold long enough to get on the price lists frequently published in the agricultural press), the powerful U.S. equipment was in a (very small) class by itself. In 1967 the company again began sourcing smaller models from Germany to complement its U.S.-made equipment and then offered the broadest range of tractors in the British market.[37] Nevertheless, while the line's quality was respected, its prices were high, and by the end of the decade, Deere still had a miniscule (unestimated) market share.

The only other entrant into the British market before the late sixties was Porsche. In 1955, 1959, and again in 1962, just before being taken over by Renault, extensive advertising campaigns were launched, but the prices offered were never particularly attractive.[38] As in the U.S. case, there was apparently an exaggerated estimate of the design's appeal by its manufacturer.

Entry from Eastern Europe

Sustained efforts by several Eastern European firms which, as is the case in the EEC, enjoy most favored nation trading status showed some signs of success in the late sixties. The Czech Zetor was launched in 1966 (and confounded skeptics by maintaining perfect parts availability after the Russian invasion), and the Polish Ursus, the Hungarian Dutra, and the Rumanian Universal (as well as the Finnish Valmet) were all being offered in 1970. The principal selling point was price—as much as 25 percent lower on some models and among them horsepower offerings covering nearly all classes of British purchase.[39] Zetor achieved a 5 percent market share in only two years.[40]

FRANCE

Initial Post-War Entry

The history of the French tractor industry in the first few years after the war is rather complex because many domestic firms considered and then modified plans for participation in the government's comprehensive scheme for rural mechanization which was directed in large part through the provision of finance. In the earliest years, the pattern of war-time armaments was pursued (and with some of the same firms) in which factories of several domestic firms divided production of similar models, although state-owned Renault produced as much as 70 percent of annual output.[41]

Massey-Harris and International Harvester had both operated implement factories in France prior to the war and were anxious to retain their role in the market. In order to do so, however, they discovered they would have to contribute to French "self-sufficiency" in tractor production, a declared aim of the Monnet Plan. When the two North American firms began manufacture in France in 1950 with considerable financial support from the government,[42] both produced machines reflecting so much engineering and marketing experience that they quickly outdistanced the inexperienced Renault operation and other domestic rivals. Their competitive strength is scarcely surprising. Until the mid-fifties, unlike the situation almost everywhere else outside of North America, gasoline was the most economical fuel in France, and the two firms simply produced virtually unmodified versions of their smallest home market models; the simplest and cheapest "lugging tractors"—without hydraulics of any kind—were in greatest demand in France as elsewhere on the Continent. Nevertheless, Ferguson entered the French market by licensing an independent manufacturer in 1953 just before the merger with Massey-Harris, and a considerable number of tractors had been imported from the United Kingdom in the previous years.[43] The government-induced change in fuel prices in 1954 in favor of diesel caught the large American firms off-guard and gave Renault enough competitive strength to pull it ahead of International Harvester in the late fifties, although 10 years later they were again neck-and-neck. Massey-Ferguson retained its lead throughout. The development of market shares in France is given in Figure 7–3.

The history of Fiat's entry into the French market is somewhat obscure. An early entrant (before 1949), M.A.P., had become M.A.P.-Someca by the early fifties, and it appears that Fiat had majority control of this company by the end of the decade. Fiat's 1966 purchase of Chrysler's 25 percent of the parent Simca Industries (a separate firm from the automobile producer after 1960) gained full control.[44] Whatever Fiat's exact degree of control in the fifties, it provided most of the tractor offerings for Someca by the end of the decade, although Italian imports were augmented by some production in France and, temporarily, by powerful Steyr imports from Austria.[45] There appears to

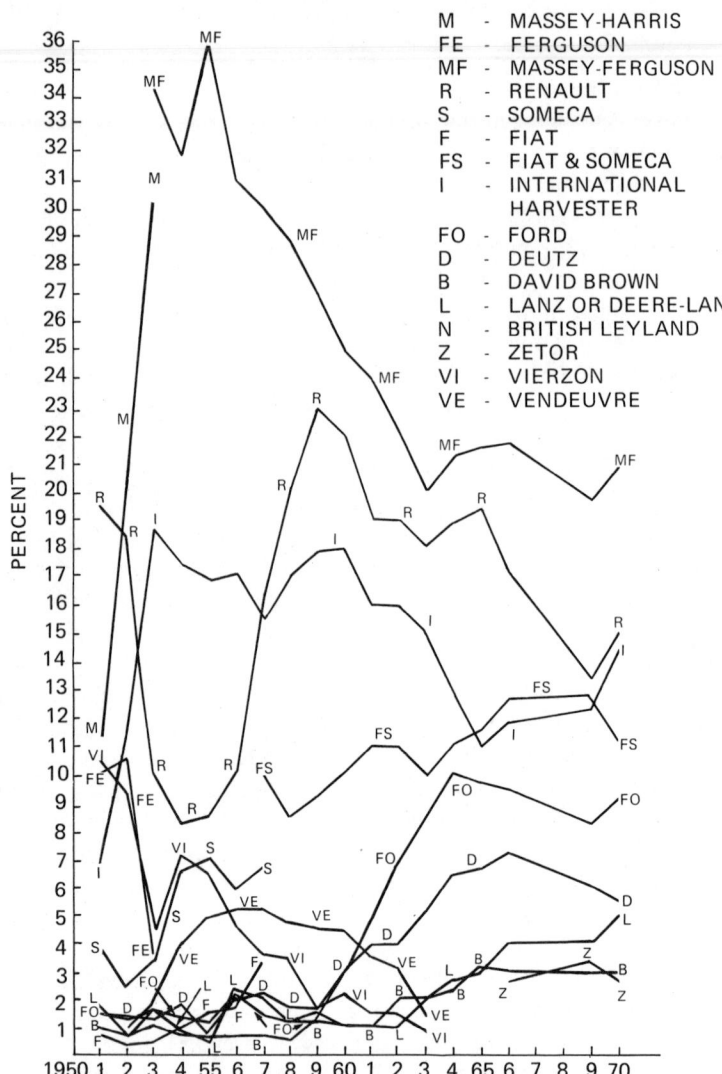

Figure 7-3. French Market Shares

Sources: 1951–1959 *L'Argus de l'Automobile,* May 29, 1958, p. 37 and May 28, 1964, p. 32. These figures are calculated on the basis of domestic production plus imports. Exports during this period, however, were modest. 1960–1965, Donaldson, Lufkin, and Jenrette, *European Equipment Industry,* p. 17. 1966–1970, C.N.E.E.M.A. *Bulletin d'Information,* various issues.

have been nothing particularly noteworthy about Fiat's strategy, but it held on to approximately the same market share after the late fifties.

Entry from Foreign Production

All other major firms have served the French market exclusively from imports, which tripled in 1956—the year intra-European trade controls were lifted.[46] Of these importers the most successful by far up to 1970 was Ford, which moved from a 1 percent share in 1958 to 10 percent in 1963, after which it fell back slightly.

There is one obvious and simple explanation for Ford's rise in France. As late as 1955, Ford's only British tractor, the "Major," was larger than over 90 percent of all tractors sold in France. By 1960, partly as a result of Ford's adding a somewhat smaller machine, the "Dexta," but mostly because of a rapid increase in the size of machine demanded in France, the company was able to serve at least a quarter of the market and, by 1964, probably more than half. (The reasons for the narrowness of Ford's range will be discussed in Chapter Eleven.) This explanation, of course, ignores the possibility that strong price competition by Ford could have been at least partly responsible for the observed shift in demand, and the fact that Ford and Massey-Ferguson relative prices for comparable models in France and Britain shows no important difference could merely indicate Massey-Ferguson's conviction that its best strategy was to match Ford's prices at every turn. It is further possible that these two firms may have moderated larger unit prices relative to those of smaller machines in order to compress worldwide demand into a narrower range of models to reduce production costs (perhaps hastening the demise of some less flexible smaller rivals along the way), but such speculation does not admit to conclusive testing. It must be remembered that running costs, and not just initial purchase price, help determine the farmer's attraction to smaller machines.

Not only was Ford the most successful participant in the French market relying wholly on imports, which came initially from the U.K. and after 1964 from Belgium, but it was also the only importer to rely solely on its own wholesale distribution system over the entire post-war period. Deere-Lanz established its own wholesaling in 1960 at the same time as the parent firm began implement manufacture in France.[47] More directly controlled distribution, a broader implement line, and extensive dealer credit led to a substantially increased market share for Deere-Lanz by the end of the decade. After the establishment of a diesel engine plant at Orléans in 1965, Deere tractors for the French market were also assembled there and engines were shipped to Germany.[48] There is no evidence that Deere consistently put downward pressure on French prices, although industry sources report that Deere, Ford, and Renault have all at one time or another taken the lead in lowering a price structure which remained essentially uniform among the major sellers.[49]

Deutz began distributing its own machines in 1969 after having expanded its implement line.[50] Other firms throughout relied almost completely on independent distributors; of these David Brown and Zetor were by far the most successful. As late as 1960, David Brown did not appear to be offering unusually low prices, but by 1966 a combination of increasing French demand for more powerful tractors and prices at least 5 percent below those of major competitors had tripled its 1 percent 1960 market share.[51] An inspection of Zetor prices reveals that significantly lower prices were not offered in France until well into the sixties at about the same time as serious attempts were made to enter the British and German markets. As elsewhere, prices of up to 25 percent lower than those of competing models were offered, and the 1.5 percent 1963 market share had doubled by 1969.[52]

Entry by Take-Over

Where a product differentiation barrier to entry based on established reputation and dealerships is important, one might expect that entry through the acquisition of a firm with a "good name" and established distribution could be attractive, and this was apparently the reason the last domestic entrant (1952), Vendeuvre, fell into the hands of Allis-Chalmers in 1960. The American firm's declared strategy was to use "Vendeuvre's extensive dealer organization to distribute Allis-Chalmers equipment in France."[53] It also declared a vague interest in obtaining "a new source for tractors and engines to be sold within the Common Market and elsewhere."[54] Material from Vendeuvre's factories was to be complemented by imports from the U.K. and the U.S. Details of the ensuing disaster are not known, but output contracted rapidly during the next several years, and the plant was closed in 1964.[55] In retrospect, it appears that Allis-Chalmers had no important advantages in the venture and should have foreseen the uphill fight it faced. It acquired a dealer network which, however well established, was serving a low-volume producer (under 4,000 units) at a time when other small firms were closing their doors. Further, Allis-Chalmers had very little previous experience in France in any product line, and its only other tractor operation outside the United States, that in Britain, was doing badly.

Case acquired tiny Vierzon in 1958 primarily for the supply of small diesel units for North American sale and only secondarily to gain a foothold in the European market,[56] but it still must have come as a blow that domestic demand dictated a 50 percent cut in production volume the first year of Case ownership. The operation stumbled along at something over 1,000 units per year for several more years. Whether the dismal history of Case-Vierzon can be assigned to excessive attention toward planned sourcing for the U.S. market, general ineptitude, or resistance to the unknown foreigner by nationals at the plant, in the selling organization, or among potential purchasers, it can be asserted that Case's entry was even less propitious than that of Allis-Chalmers. Case had no European or other foreign experience. Money was lost on the

French operation in every year until 1963,[57] and production was suspended in 1966, although several hundred large Case tractors imported from the United States continued to be sold annually.[58]

GERMANY

Tractor manufacture in Germany has a long history including pre-war contributions to design advance by Deutz, Hanomag, and Lanz, but production until the mid-thirties was never more than a few thousand units a year. The subsequent diversion of labor to preparation for war provided impetus for government support of agricultural mechanization, although, during the war itself, tractor production was subject to serious interruptions from aerial bomdardment.[59]

Initial Post-War Entry

Post-war production of tractors began under conditions of physical destruction, the division of previously unified markets (within the West as well as from the East), and material shortages. There were many domestic entrants, however, lured by the desire to engage in a kind of machinery production which would be well received by the allied administrators. The spirit of the "Morgenthau Plan" was thought to prevail in their ranks, and all contributions to an increase in the relative strength of the agricultural sector of the economy were presumed to be welcome.[60] Domestic sales tripled in 1949 after all restrictions were lifted on the production and acquisition of agricultural machines. One estimate put the number of companies producing wheeled tractors in 1951 at 32, and they accounted for 90 different models with 54 different engines, although domestic production was only 80,000 units.[61] The development of market shares in Germany from 1956, when data first became available, is shown in Figure 7–4.

Only two domestic firms entered after 1950. Hatz began selling a standard line of agricultural tractors in 1957, never attained 1 percent of the market, and gave up in the mid-sixties. Gutbrod entered in 1964, but attained something over a 1 percent share by concentrating on a very small machine of unusual design.[62]

Unlike the situation in France, tractor imports were decontrolled very early after the war in Germany, and tariffs were low, but German demand was for diesel power, and for this reason and perhaps also for marketing impact, International Harvester began producing a small diesel model in 1951—immediately upon the reconstruction of its pre-war implement facility.[63] The alternative would have been to import from Britain, but there was probably considerable uncertainty about what kind of machine would prove to be most suited to small German farms, and perhaps also a fear of market resistance to equipment imported from the victorious country. In the event, the decision was a sound one, and it marked the beginning of a sustained success by International Harvester

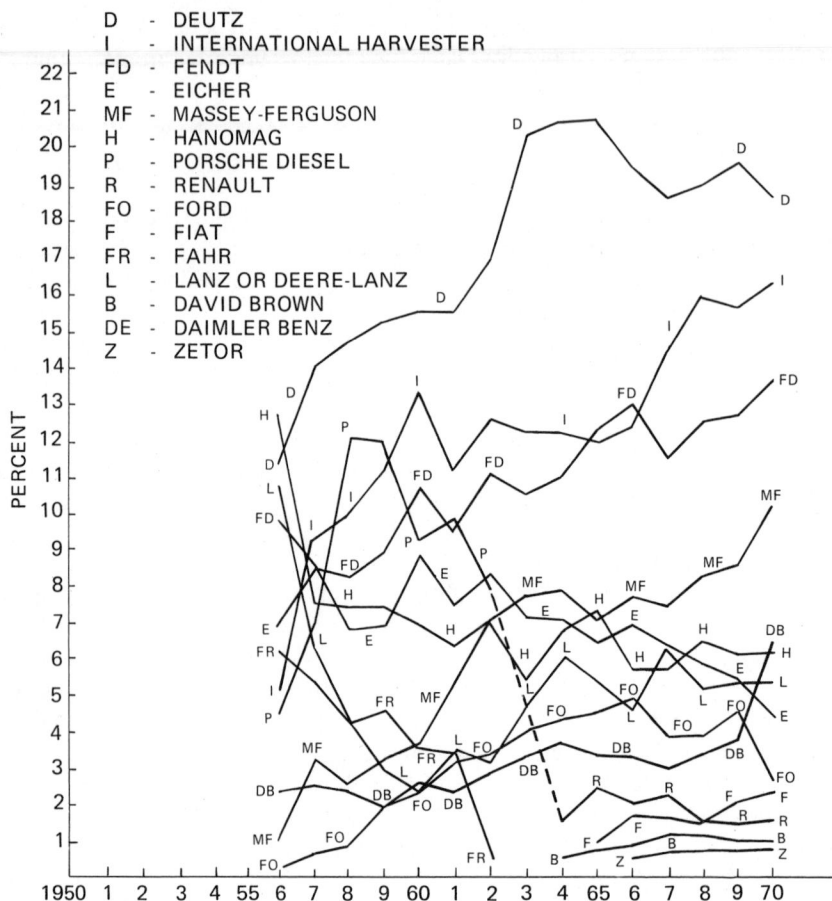

Figure 7-4. German Market Shares

Source: Registration figures from *Kraftfahrt Bundesamt* cited in *Landsmaschinen Markt,* October, 1969, p. 1076, and *Frankfurter Allgemeine Zeitung,* December 2, 1970.

which had achieved second place and a 17 percent market share by 1970. Until the late sixties, when production was coordinated with French and British plants on a common line of tractors, the German facility concentrated independently on designs advertised as "in Deutschland entwickelt und gebaut" (developed and built in Germany), while at the same time never attempting to disguise its North American origins. As in all other foreign markets, it used only the same names used at home: "McCormick," "International," and "Farmall."

Massey-Harris considered production in Germany at the same time as did

International Harvester, but its German facilities had been totally devastated and, because of capital shortages, it began to serve Germany from France, a practice which was continued indefinitely.[64] Although it may well have suffered some marketing disadvantages from so doing, its market share expanded smoothly and impressively from only 1 percent in 1956 to over 10 percent 14 years later. It was presumably the massive devaluations of the franc which accounted for the firm's ability to double its 3 percent 1959 share in Germany by 1963, although whatever downward price pressure Massey-Ferguson put on the German market structure was matched within a period of time too brief to be recorded in the exiguous price data available. What is known is that there have been no sustained and substantial price differences for similar models of the major participants in the German market.

Deere's Entry

Deere entered the German market in 1956 by acquiring the family firm of Heinrich Lanz, and its indifferent record thereafter can be partially explained by the fact that, although Lanz's market share in 1956 was 11 percent, its tractor offerings, like those of Deere at home, were in great need of revision. A detailed examination of the reasons for Deere's entering into manufacture outside of North America will be presented in Chapter Eleven. Unlike the other three of the "big four" it had little pre-war experience or stake abroad, but the German move was in part an attempt to get in on the ground floor of the EEC, and the fact that Lanz was the only major Continental firm producing its own implements, including a combine,[65] may have helped Deere to decide which firm to buy. Full-line marketing was the only kind with which Deere had any experience, and Lanz's failing fortunes may have made the purchase price attractive.

Deere immediately attempted to exploit its reputation as a principal capital goods supplier to the U.S. corn belt with its worldwide image as an area of unsurpassed productivity. Despite Lanz's third-place market share in Germany and long-standing reputation in the German industry, the famous U.S. corporate emblem and green and yellow color scheme were immediately adopted, and the name of the firm became Deere-Lanz. Predominantly German management let the firm fall to a 2.4 percent market share in 1960, partly because of product difficulties with the new tractor line introduced that year, and only with an infusion of management from the U.S. did the situation improve.[66]

Despite its less than spectacular performance subsequently, evidence that Deere was substantially correct in attempting to rely on its North American identification in marketing seems to be confirmed by the behavior of some European firms. The German family firm, Fendt, called one of its most popular models in the sixties "Farmer," and Porsche earlier called one of its models "Junior." In 1965 the Italian firm, Same, dubbed one of its models "Atlanta." While the U.S. automobile industry has often attempted to give "class" to its offerings by giving them European names (sometimes misspelled and usually mispronounced), it strains one's imagination to consider a tractor on a farm in Iowa bearing the name "Bauer," "Fermier," or "Agricoltore." The apparent asymmetry, as hypothesized in Chapter Four, is almost certainly due to the image of high U.S. productivity in agriculture (and its attendant relatively high incomes) which undoubtedly serves to enhance at least somewhat the European user's self-esteem in much the same way as does driving a "Coupe de Ville" by an American parvenu. By contrast, the (not altogether outdated) image of the Continental farmer in the eyes of his North American counterpart is that of a backward peasant.

Entry from Foreign Production

The possible special advantages that North American-based firms might have in dealing with customers otherwise resistant to unfamiliar foreign machinery is seen by contrasting Massey-Ferguson's experience with that of Renault which entered Germany in 1963.

Renault, which appears to have been partly motivated by a desire to put more of its staff on full-time work (tractor production was 19,000 in 1961, but only 15,000 in 1962),[67] simply bought the distribution facilities of Porsche-Diesel and M.A.N., firms which had an extensive and largely exclusive dealer network of 1,500 outlets.[68] Tractor production by the parent Mannesman, which had failed to establish profitable volume in previous years, despite the absorption of several small firms, was then abandoned. Although adoption of the Renault line would have been the most obvious course for the stranded dealers, they deserted in droves, and Renault's hoped-for 12 to 17 percent of the German market had evaporated to 1.7 percent in 1964, the first full year of the newcomer's participation. In several following years Renault's share rose above 2 percent but slipped to 1.6 percent in 1970. How much of the disaster can be attributed to poor management by Renault and the possible shortcomings of its products and how much assigned to demoralization engendered by what dealers called "going over to the French camp" (especially keeping in mind why Renault was nationalized) cannot be known, but contemporary accounts suggest that the latter factor was important and must have been largely a judgment about probable customer reaction to the new line.[69]

Besides Massey-Ferguson and Renault, the only firms serving the German market from imports with non-negligible market shares at any point during the period were Ford, Fiat, David Brown, and Zetor; and, of these, only Ford was participating at all before the mid-sixties. Ford's product line limitations were even more important in penetrating the German market than they were in France; as late as 1964, its line covered less than half the market, and there were only about 70 dealers with whom Ford dealt directly.[70] As part of a strategy to improve its German position, Ford launched a form of price war aimed at shaking up dealer-manufacturer relations in the industry. German dealer discounts from suggested retail price were 32 to 37 percent by comparison with 16 to 27 percent elsewhere, and this margin offered so much latitude that a typical dealer often didn't know where to begin an offer to a prospective customer. In June of 1967, Ford revised its list prices downwards to allow a margin of about 14 percent, and most other firms followed suit.[71]

Not surprisingly, there was a section of the retail trade which responded most unfavorably to the reduced latitude introduced by the manufacturers and saw it as a way of forcing smaller dealers out of business.[72] According to information received by the writer from the German manufacturers' association, the new discount level was later judged to be too low by most manufacturers, including Ford; and, in the resulting readjustments, "recommended retail prices" which, if in existence, had to be reported to the *Kartellamt,* were simply abandoned.[73]

Fiat began selling through independent distributors in Germany in the mid-sixties, pursuing an avowedly low price policy,[74] and it moved from .9 percent of the market in 1965 to 2.5 percent in 1970. Unfortunately, because German price information by firm is virtually non-existent except for RCFM data for 1966, comments about price competitiveness must necessarily be tentative. Based on that one year, however, Ford, Fiat, and David Brown appear to have been selling somewhat more cheaply (perhaps 5 percent) than most of their competitors.[75] David Brown established its own distribution system in Germany in 1964,[76] the same year it had a large enough market share, .6 percent, to show up in the national registration statistics, and this rose to about 1 percent in 1970.

The Zetor line which first surfaced in published market shares in 1966 and still had less than 1 percent in 1970 was probably sold at price differentials similar to those in France and Britain, but no hard data are available. That the future of Eastern European tractors in Germany was regarded as promising, however, is evidenced by the construction in 1970 of a large German assembly plant for the Fiat-designed Rumanian Universal tractor. A small German specialist manufacturer was to market the machine in Germany and neighboring countries, and the avowed purpose of the venture was to sell completely modern tractors at "extremely low prices."[77]

ITALY

Foreign Domination and Its Decline

Foreign tractors, especially those from the United Kingdom, dominated the Italian scene in the first few post-war years because the foreign producers could undersell their tiny domestic competitors despite a 40 percent tariff. Further, the Italian government chose not to make it even more difficult for the foreign sellers, because it was necessary for a number of years to allow available foreign goods access to the Italian market to balance against Italian exports of wine and fruits; German and British tractors served this function.[78] The Fanfani Plan of 1952 put outsiders at a considerable disadvantage, but, while the David Brown and Ferguson shares dropped sharply, Ford's position was maintained, probably because of the importance of Ford's Italian automobile sales subsidiary through which some of the tractors were sold. Ford suffered about the same size disadvantage in Italy as in France, although not as much as in Germany; the rugged Italian terrain partly compensated for the small size of landholdings, and, as early as 1955, average wheeled tractor power was nearly 30 h.p. Ford did not quote prices as low as those of the leading Italian firms, but it came closer than other importing firms over the entire period.[79] The development of Italian market shares is shown in Figure 7-5.

In the earliest post-war years, the government played a considerable role in the domestic industry through the Institutio per la Riconstruzione (I.R.I.) which operated Ansaldo, Motomecannica, and O.T.O. These firms together produced 26 percent of all domestically sold wheeled units in 1951,[80] and there were a multitude of smaller firms. By the mid-fifties, the situation had changed dramatically. Many small firms survived (although there were virtually no new entrants), but the share of the I.R.I. firms had dropped to only 4 percent in 1956, and Fiat had emerged as the industry leader, increasing its share in nearly every subsequent year.[81]

By 1956, the major foreign producers, none of which had a manufacturing facility in the country, were serving only 37 percent of the market by contrast with 55 percent in 1950; their role fell yet further to 30 percent in 1960 and to 14 percent in 1969.

It is not altogether clear why no foreign firm entered the Italian market during its initial phase in the fifties, but an important reason was probably that total Italian home demand remained below 20,000 units as late as 1960, by which time tariff-free access to the Italian market (and presumably the ultimate demise of the Fanfani Plan) appeared to be in the offing for firms with other EEC plants.

Massey-Ferguson's Entry

The family firm, Landini, which before the war and in the early post-war years was the largest seller, relied during the fifties on machines powered by engines of an out-dated design. The firm turned to Perkins of Peterborough (U.K.) in 1958 to obtain licensing for the production of a modern line of diesel engines, but even then foresaw the ultimate weakness of its position and welcomed a purchase offer from Massey-Ferguson in 1959 (the latter had acquired Perkins late in 1958). Massey-Ferguson's declared aims in acquiring Landini were 1) to become established before EEC tariff cuts became effective; 2) to overcome the bias of the Fanfani finance plan; 3) to gain an established dealer organization; and 4) to gain a crawler technology. Within Italy Massey-Ferguson then marketed two lines: Landini tractors and Massey-Ferguson units imported

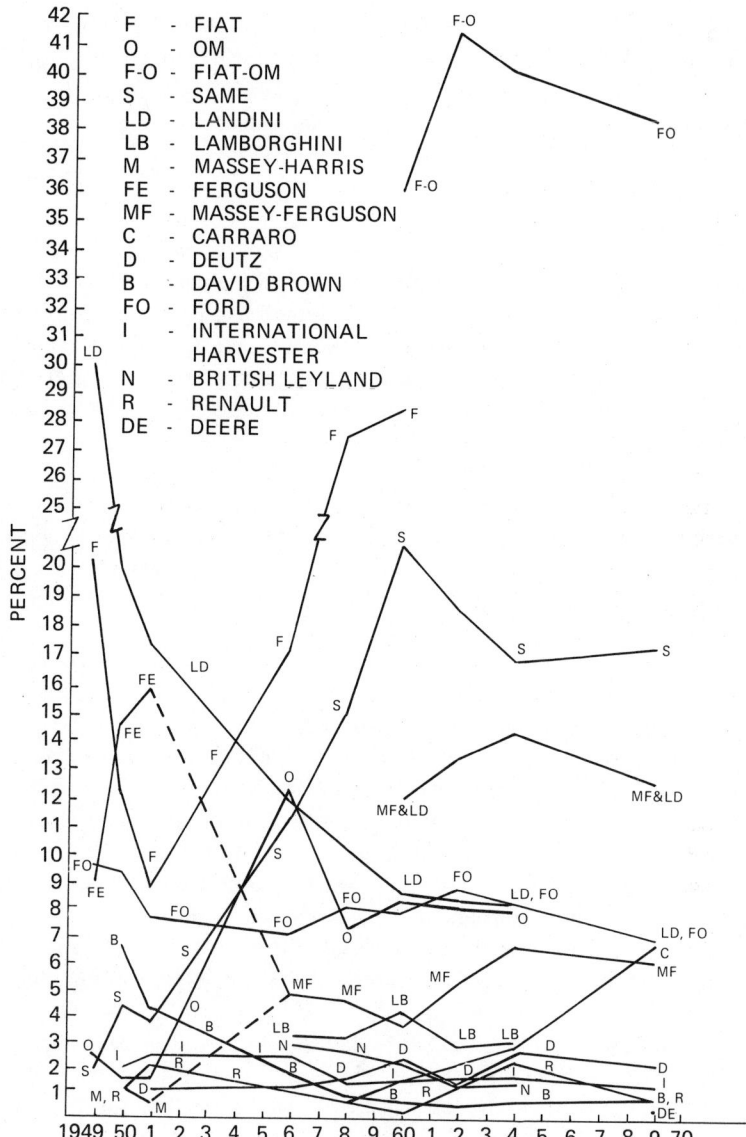

Figure 7–5. Italian Market Shares

Source: Utenti Motori Agricole, *La Meccanizzazione Agricola in Italia,* various years; selected 1969 estimates from the Ford Motor Company.

from France and Britain. Experience gained with the new operation and the ability to quote lower prices as Italy's formidable tariff barriers were lowered for equipment coming from France led to a considerable increase in the sale of Massey-Ferguson machines which were sold independently from Landini equipment.[82]

AUSTRALIA

Although Australian manufacture of tractors beyond the merest handful did not begin until after the war, as early as 1938, International Harvester had decided that the Australian government's determination to have a domestic implement industry (through heavy protection) meant that International Harvester would have to build implements there to complement its U.S.-made tractors or tractor sales might suffer.[83]

Initial Post-War Entry

Post-war discrimination against the dollar area which had previously supplied more than three-quarters of tractors sold and government subsidies for tractor manufacture induced five firms to begin production: McDonald, Chamberlain, Howard, and K.L. (all Australian-owned firms), and International Harvester, which began to make tractors at its already-built implement facility. All but Chamberlain and International Harvester had dropped out by 1955, and the former was not producing profitably. Although Chamberlain was obtaining financial underwriting from the industrialization-minded Western Australian state government, by 1958 the firm was under bank control.[84] Furthermore, International Harvester began decreasing its Australian production in 1955 and 1956 and increasing importation from the United Kingdom,[85] a practice which the leading firms in the market, Massey-Ferguson (initially Ferguson) and Ford practiced exclusively throughout.

Australia is the market in the study about which least is known about firm shares. The Ferguson share was 17 percent in 1948 and 34 percent in 1952, and Massey-Harris had an 8 percent share in 1948 and 6 percent in 1952; their combined share was 34 percent in 1955.[86] In 1956 Ford's advertising implied an approximately 25 percent market share.[87] At the same time, International Harvester's share was probably at least 15 percent, giving the top three firms about 75 percent of the market.

One investigator, basing his estimate on an interview with a firm executive, claims that the top three firms in 1962 were still Ford, International Harvester, and Massey-Ferguson.[88] This source implies that Chamberlain's output was such that it could have had no more than 10 percent of the market.[89] Brash's study of U.S. investment in Australia identifies the same three firms as leading in the early to mid-sixties. A trade source in 1969 confirmed their continued predominance and also reported that they then supplied about 70 percent of all tractors sold in Australia. Another source suggests that Massey-Ferguson was the largest seller in 1969.[90]

A new subsidy schedule introduced by the government in 1959 enabled Chamberlain and International Harvester to expand Australian production somewhat; average national production from 1952 to 1958 was under 3,000 units, while the average over the period from 1959 to 1970 was over 4,000 units;[91] subsidies were raised from a maximum of under 25 percent of wholesale price to over 35 percent in the later period.[92] The subsidy figures given are for 90 percent or more Australian content, but in fact little of the production would have qualified for that rate, because Chamberlain relied wholly on imported engines (about a quarter of total cost), and most of International Harvester's offerings over the period also used imported engines (and many whole machines were imported as well).

David Brown and British Leyland (and predecessor companies) served the market continuously after the war, although neither appears ever to have attained more than a few percent of the market. Price information is scarce, but in the mid-sixties British Leyland's prices were not unusual while those of David Brown were about 5 percent lower than those of International Harvester, Ford, and Massey-Ferguson and were cheaper than any line except Fiat.[93] David Brown claimed in the late sixties that its Australian market share had increased "four or five times" over the preceding 10 years and that it had moved into company-owned wholesaling *pari passu* with a certain critical minimum of actual or anticipated sales in a given area.[94]

Entry in the Fifties and Sixties
Because some German tractors were sold in Australia before the war, it is not surprising that Lanz and Deutz attempted to sell there in the fifties, as did Renault.[95] All three firms apparently gave up well before 1960. By contrast, Fiat was sold from the early fifties with increasing success; it was apparently favored with an excellent distributor for Queensland and New South Wales, the richest farming areas, and the tractor line was sold at prices considerably below those of the largest sellers. Fiat's share moved from 1 or 2 percent in the mid-fifties to between 5 and 10 percent by the late sixties when Fiat may have moved ahead of Chamberlain as the fourth largest seller in the market.[96] Accurate price figures for the mid-sixties suggest that Fiat tractors were selling for at least 10 percent under those of the three predominant firms, which sold comparable machines differing from each other in price by only a few Australian pounds.[97]

When dollar discrimination was lifted in 1957, the North American full-line firms with only U.S. production began to re-enter the market, but all were able to sell only their largest models, and none attained a substantial market share.[98]

Czech Zetors, entering Australia on the same terms as all other machines except those from Britain, began being sold in the late sixties at extremely attractive prices, but nothing is known of their market penetration.

PATTERNS OF ENTRY

In reviewing the course of entry over time, a few generalizations are possible.

The Importance of Domestic Fabrication

Despite the supposed advantage of domestic fabrication in producer goods, its importance is difficult to identify in the tractor industry. While it may confer a minor marketing advantage, other factors have usually overwhelmed it. The success of imports in all markets appears to relate more to the image of country of origin (either in manufacturing or agriculture) or of the firm itself than to foreignness per se. Although production in North America for sales there (the virtually universal practice of all major firms throughout except Ford and Massey-Ferguson) may have been considered by some to confer a marketing advantage, the practice is inextricably linked with the problem of moving production to lower cost sources of supply, an important issue which will be examined in Chapter Eleven. Ford in the mid-sixties began British sourcing for nearly all components for many models sold in North America with no apparent concern about marketing impact (see pp. 173–174).

Importation of whole tractors into North America did remain minor, but this appears to be the result of decisions internal to the firms dominating that market and well protected from *all* new entrants by the enormous product differentiation advantages they held. The minor role of importation into Britain and Italy appears almost exclusively due to the inability of outsiders to match internal price levels (see Chapter Ten); there is no evidence that foreign production per se was of particular importance. The unimportance of the consideration when other factors are propitious is illustrated by the experience of France. Not only was about half of third-place Fiat's product line imported throughout, but the spectacular growth of the "second four" (Ford, Deutz, Deere-Lanz, and David Brown) in France's top eight to nearly a quarter of the market was accomplished wholly through imports. Chapter Eleven will examine Ford's establishment of an EEC production facility, but, if marketing considerations had been critical, this plant would certainly have been located in a major national agricultural market rather than at the site of a disused automobile assembly plant in Belgium. Furthermore, an examination of the German market in the sixties reveals Massey-Ferguson, importing from France, to be the most impressive major participant measured by the increase in market share. While Deutz and International Harvester each increased their share by about 4 percentage points during the sixties for an increase of about 30 percent apiece, Massey-Ferguson increased its share by six points, a 150 percent increase. It can be argued, of course, that intra-EEC trade isn't international in the usual sense of the term despite the fact that, formally, goods "must cross oceans or national boundaries,"[99] yet as late as 1968, France briefly imposed quantitative import restrictions on tractors [100] (though there were presumably internal inventories of both parts and whole machines).

In Australia, whatever marketing advantage might otherwise have attached to International Harvester's domestic fabrication was overwhelmed by the appeal of its Ferguson-system competitors. The pattern of imports into the various markets over time is shown in Table 7-2.

The Paucity of Production Extensions

With economies of scale as important as they are in the tractor industry, virtually stagnant unit demand since the fifties, and no significant product adaptation necessary for individual markets, it is not surprising that all production extensions (beyond the most minor assembly operations) fall into three categories. First, there was the move to Britain and the Continent by the major firms in order to avoid being indefinitely shut out from rapidly growing markets.[101]

Table 7-2. Import Shares, Selected Markets (Percentage)

	North America	United Kingdom	France	Germany	Italy	Australia
1950	nil	nil	59	b	55	100
1951	nil	nil	42	b	45	100
1952	1	nil	32	b	59	81
1953	1	nil	24	b	74	86
1954	4	nil	18	b	38	83
1955	2	nil	11	1	38	83
1956	3	nil	20	3	34	88
1957	3	nil	20	5	25	78
1958	2	nil	15	5	25	102[c]
1959	5	nil	15	7	26	76
1960	4	nil	24	8	24	88
1961	4	nil	30	11	20	95
1962	6	nil	34	13	27	62
1963	5	nil	41	16	26	90
1964	5	nil	40	16	27	67
1965	3	nil	42	20	–	105[c]
1966	5	a	41	21	24	65
1967	4	a	39	13	24	64
1968	4	a	43	18	–	75
1969	5	a	43	19	14	72
1970	5	5	43	–	–	78

[a]Under 5 percent.
[b]Unrecorded but probably less than 1 percent.
[c]Percentage exceeding 100 percent due to anticipatory inventories.
Source: See Figures 11-1 and 11-2.

Second, there was the expansion of some facilities in conjunction with the coordination of production between Britain and the Continent undertaken by International Harvester, Massey-Ferguson, and Ford in the sixties, which was a move to minimize delivered costs given the apparent finality of Britain's non-entrance into the Common Market and the 18 percent external EEC tariff (see Chapter Eleven).

Third, there were the acquisitions of production facilities as an integral part of the takeover of existing firms in established markets by tardy foreign entrants. In three of the four cases discussed, the principal desideratum was established distribution and not either tractor design or production facilities. Both product design and plant demanded substantial modernization in the cases of Deere-Lanz, Massey-Ferguson-Landini, and Allis-Chalmers-Vendeuvre. The other venture, Case-Vierzon, appears to have been based on a complete misreading of the design and production capability of the acquired firm (see p. 120). Furthermore, it should be stressed that these acquisitions provided more than one production point within the EEC only for Massey-Ferguson.

The Role of the Largest Firms

The pattern of international entry certainly confirms the hypothesis that the largest firms will tend to dominate the activity,[102] although this outcome was modified by the confirmation of another familiar hypothesis: that firms invest to protect established markets. Massey-Harris was a firm of only modest size before World War II by comparison with most of its North American competitors, but very early in the century, partly in an attempt to avoid the competition of U.S. firms, it vigorously sought out markets around the world. Nearly half the firm's output, mostly harvesting machinery, was sold outside of North America as early as 1908, and in 1939 the figure was around 40 percent.[103] Its very active post-war strategy outside of the U.S. and Canada can be explained by the critical importance of foreign markets, despite the fact that it was perhaps the smallest North American-based manufacturer of wheeled tractors in the earliest post-war years. International Harvester's foreign role was large and well established before the turn of the century, with 40 percent of its sales outside of the United States as early as 1912. The immediate pre-World War II stake was about 20 percent.[104]

John Deere, the second largest U.S. firm, had far less pre-war experience than did International Harvester. Deere became a large and broadly based implement firm only a few years before World War I and subsequently showed little interest in overseas expansion; instead, it successfully concentrated its attention during the twenties and thirties on expanding its share of the home market. Only after the post-war boom had subsided, for reasons discussed in Chapter Eleven, did Deere realize the critical importance of overseas activities, and this realization led to the purchase of Lanz. Of the other North American full-line firms, only Allis-Chalmers, the fourth largest in U.S. farm machinery sales at the war's end, became immediately active in post-war overseas markets by beginning tractor production in the U.K. The reason for this activity may well have been the pre-war experience of the firm in overseas markets, only a modest amount of which was in farm machinery, however. Allis-Chalmers' 14 percent of sales outside of North America in 1947 was approximately the same as in the pre-war years and was far higher than that of Deere, Oliver, Case, or Minneapolis-Moline. Allis-Chalmers' entry may well represent the action of a firm which because of its other activities, had established "international decision horizons." [105]

The review of non-North American-based firms can be more brief. The tractor operation of the British Ford company (which was not a wholly-owned subsidiary of the U.S. firm at the time) had during the thirties sold throughout the Empire, [106] and at the war's end, it was the largest non-North American tractor-maker. Ford was quickly challenged by Ferguson, and the latter's British operation was larger than Ford's when Ferguson acquired its French licensee in 1953.

The other British firms with overseas activities, David Brown and Nuffield, probably ranked after Ford and Ferguson as the largest non-North American firms in the immediate post-war period. They were subsequently surpassed by Deutz and Renault, temporarily in the late fifties by Porsche, and later by Fiat. [107] This virtually exhausts the set of non-North American non-Communist international entrants, just as it exhausts the set of large non-Communist firms in the tractor industry outside of North American at virtually every point in time.

Product Line and the Success of Entry

A consideration obviously bearing on the success of entry, once attempted, was the type of equipment offered. Some limitations of tractor offerings have already been outlined. Ford suffered on the Continent over much of the period from the high power of its tractors, as did David Brown and British Leyland. In North America, by contrast, Ford, David Brown, and British Leyland remained limited to what in that market were small and medium units. In addition, all three of these firms suffered in North America and increasingly on the Continent as well as from the absence of a complete line of equipment to sell with their tractors, a factor to be discussed in detail in the following chapter. Fiat's range of tractors was comprehensive by Continental standards throughout, but the firm began most of its foreign activities later than the other large firms, and it, too, presumably began to suffer in France and in Australia by the end of the period from the absence of a full line.

The Role of Independent Wholesalers

Those firms entering with modest sales expectations or an incomplete product line have relied heavily on independent distributors. Considering the problems facing the foreign entrant outlined in Chapter One, the use of the independent wholesaler clearly obviates one problem and diminishes another. The enormous cost of acquiring information about the market to be entered is for the most part eliminated by the use of an independent distributor, while the distributor's knowledge of sources for the packaging of equipment with tractors at the wholesale level might substantially increase the tractor line's attractiveness to potential dealers by lowering transactions costs, although the resulting line is almost certain to be less complete than that offered by the full-line firms. On the other hand, the drawbacks to the use of independent distributors are also obvious. For a firm which either offers a substantial range of equipment other than tractors (whether the line is "full" by some criterion or not), or one which regards the provision of dealer credit (to be discussed in the following chapter) as a desirable promotional device, the possibility of conflict arises. There may well be a somewhat uncertain commitment to the maximum promotion of the line of the importing firm by the distributor, and, even where the distributor provides dealer credit, it might well not be to the extent regarded as desirable by the manufacturer.

Weighing the issues, it is scarcely surprising that full-line firms entering markets above a certain size seem to insist upon controlling their own distribution to the dealer level and that tractor makers alone may sometimes find it desirable. The size of the market is clearly of great importance, because many of the costs of gaining the information necessary to serve a market are "fixed in the sense that they do not vary proportionately with the amount of resources that the firm might stake abroad."[108]

Some critical mass of activity, whether actual or anticipated sales volume in tractors (David Brown in Germany in 1964 and in Australia in the mid-sixties), or the broadening of the firm's implement offerings (Deere in France in 1960 and in Britain in 1961, and Deutz in France in 1969) will induce forward integration which, ceteris paribus, should be expected to be followed by increased market penetration (as it clearly was in the cases of Brown in Australia and Germany and Deere in France). That the U.S. full-line firms resist compromise is illustrated by the behavior of Case and Allis-Chalmers in Australia where distribution is controlled to the dealer level in the most densely farmed parts of the country (see p. 131).

There is bound to be a lot of "noise" in the relationship just stated, and the crossover point in sales volume will vary even for the same firm by country, as would be expected given the great variation in the quality of independent distributorships. David Brown, for example, continued to use an independent distributor of long-standing in France in 1970, although it set up its own distribution in Germany six years earlier with a sales volume only a quarter as large, and both David Brown and British Leyland found the established relations of their U.S. distributors with complementary implement producers and dealers far more valuable than any additional control over dealers which might have been possible with direct contact.

The possible disadvantages just suggested for the use of independent whole-salers would appear to have less to be weighed against them when the producing firm is *domestic,* and the demise of such wholesaling, to be recorded in the following chapter, should come as no surprise.

Entry and Price Strategy

Price policy and competition admit to only limited discussion because of the paucity of accurate information, but what evidence there is suggests that, al-though in every market the prices of the largest sellers have been very similar, entering firms often attempt to sell at lower prices, and in some instances these prices are not matched. It appears that where David Brown and British Leyland sold at low prices, they never induced a matching response. Fiat sold noticeably more cheaply in the two markets where it was a minor participant (Germany and Australia) than where it was one of the major firms (Italy and France), and Ford appeared to sell at noticeably lower prices in the market in the study where it had the smallest share (Germany). It is probably significant that none of these firms was offering a complete implement range in any of the markets in which it was selling at a noticeable discount. The lack of a complete line presumably most diluted the perceived long-run threat in those markets already dominated by full-line firms, and in North America and Australia the threat would be further lessened by the limited power range of tractors offered (the Fiat range sold in Australia was "dense" but lacking in high power units). Either one or both of these considerations could be expected to so vitiate the appeal to dealers and customers that the established firms would allow the other tractor lines to sell at a considerable discount, despite a growing market share. By contrast, there is no record of Deere ever selling at a discount in any of the markets in which it was attempting to increase its market share in the fifties and sixties. It may be that the firm seldom seriously tried; its market share at home grew despite, if anything, prices on average having been slightly higher than those of its main competitors. What is probable is that its full line and marketing expertise would have posed such a long-run threat that any downward price pressure would be immediately matched by its rivals.

The heaviest price competition came from the Eastern European entrants. In the late sixties, machines from Czechoslovakia, Rumania, and Poland were sell-ing in several of the markets at suggested retail prices up to 25 percent lower than competing models. It appears that the major sellers in all of the markets are willing to see a large price differential between their equipment and that from Eastern Europe, because they probably believe that the stigma of origin blunts the thrust of the Eastern European attack. The major firms' reasoning is presumably that in addition to the difficulties which have historically attended state trading (and which Western competitors would undoubtedly explain to prospective customers lest they be unaware), there is probably a special stigma derived from the notoriously low agricultural productivity in the Communist countries and suspicions about the quality of goods manufactured there. In addition, anti-Communist sentiment is strong in the rural sector in every market. Nevertheless, the increase in penetration of this equipment, especially Zetor tractors from the Skoda works, one of the major industrial complexes in Europe, indicates that ultimately price differentials such as those prevailing in the late sixties could lead to substantial market shares for the Eastern Europeans.

Entry and Oligopolistic Interdependence

It is difficult to find evidence of entry behavior which might be traced to perceptions of symmetry among the firms such as clear spheres of influence, reactive thrusts resulting from a breakdown of understanding about them, or any other kind of conscious division of markets. This is understandable given the complex imbalance of market strengths by the major world competitors in the earliest part of the period; any subsequent symmetrical mutually regarding behavior would be difficult to predict. Although International Harvester and Massey-Harris had vast overseas markets, including implement manufacturing facilities, prior to World War II, and Ford's British marketing base had been used to penetrate much of the Empire, Deere and the smaller U.S. full-line firms faced a very different post-war situation. For these firms, farm machinery exports had been marginal before the war, and foreign experience was very limited. For Continental tractor makers, pre-war experience was modest and, except for a few German exports, almost completely within the home market. There may very well have been some symmetrical mutually regarding entry behavior among implement markets before the war by Massey-Harris and International Harvester, but even between these two firms, no reciprocal post-war entry activity (or spheres of influence) can be discerned. While both firms were anxious to protect their investments and extend them into tractor manufacturing on the Continent, Massey-Harris found it necessary to abandon Germany and, although it did so with reluctance, its corporate biography fails to indicate that its Continental fears and hopes were more related to the actions of International Harvester than to those of several of its other rivals in Europe.[109]

It must be stressed that what the evidence adduced supports is the hypothesis that the action of a given firm in entering a new market, or in considering entry by others into a market in which the firm is already participating, appears to be a function of factors bearing directly on that market and not to factors elsewhere in the industry. Nevertheless, one of the factors directly connected with any market is the threat of entry into it and the competitive positions of the best-placed potential entrants. Here, as the following chapter will suggest, there has undoubtedly been considerable mutual dependence recognized in the sense that significant potential entrants are seen clearly as a small set of well-identified firms and "oligopolistic interdependence of crossed conjectures"[110] obtains between members of the small group within the market and members of the small group which might participate.

In short, the tractor industry has produced no direct or indirect evidence of (for example) A's taking account of B's possible entry into A's home market when it is deciding whether or not to enter B's home market or of an intrepid move by A followed by what appears to be retaliation by B. This should be contrasted, however, with the possible identification of B as being among the most likely entrants into A's market both before and after A's arrival on B's home ground.

In addition to the kind of mutual dependence recognized concerning entry just outlined, the following chapter and Chapter Eleven will discuss two other areas in which international mutually regarding behavior has been very important: the achievement of scale economies and the development of a broad range of farm machinery where it was not initially offered.

Chapter Eight

Competitive Behavior and Structural Change

The previous chapter investigated patterns of entry and some of the subsequent strategy of the entering firms. In this chapter we explore the reaction of established firms and the overall development of the markets. How have competitive pressures affected the development of distribution systems and product lines? Leaving aside physical changes in tractors and advertising (which will be dealt with in the following chapter), what other competitive or coordinated behavior can be observed in the various markets? How significant has been the role of cooperative activity in the production and distribution of tractors? How has structural change proceeded in the various markets and overall?

NORTH AMERICA

Chapter Four stressed the extent to which tractors became increasingly similar over the post-war period, and Chapter Five emphasized the rather stable pattern of distribution which developed in the North American market in the pre-war years and continued over the period under review. Nevertheless, a discussion of the post-war North American market would be incomplete without a review of the relative competitive strengths which the established firms brought to the post-war market. In particular, it is commonplace in the industry to regard the loyalty of customers to Deere and Company *and* to its dealers to be higher than to any other brand. This was recognized long before Deere overtook International Harvester as the largest seller of farm machinery in North America. There appear to have been two main reasons for this fierce loyalty; one was the product line itself. As noted in Chapter Four, throughout the inter-war period (and indeed until 1960) Deere produced a highly differentiated line of tractors—essentially refined versions of the two-cylinder "Model D" introduced in 1923. The company's cumulative reputation for the dependability of its products during this period may be due in large part to the experience it gained by concentrating on an essentially stable design. The Deere line featured a low-speed engine which was cheap to operate, flexible in fuel type, and constructed in such a way that the farmer could do much of the repair work himself. On the other hand, the machines were noisy and emitted particularly noxious fumes.

Complementing the increasing dependability of the tractor line was Deere's early and continuous emphasis on dealer consultation and support. Deere always eschewed "curbstone" fly-by-night dealing with which many of the major manufacturers were associated in the inter-war period, and even those farmers who didn't favor the particular features of the Deere tractor line came to respect the assurance of minimum "downtime" provided by the line and its high-quality dealer force. This product differentiation is undoubtedly responsible for Deere's relative imperviousness to innovations in tractor design prior to its own change of product line.

Deere's position stands in sharp contrast to that of the firm which led in North American farm machinery sales until the late fifties, International Harvester. Although the "Farmall" of 1925 took rural America by storm, in subsequent years Harvester devoted its attention to a number of different tractor designs and appears to have made considerably less effort than Deere to maintain good relations with dealers. The FTC Reports of 1938 and 1948 recorded a far greater volume of dealer complaints (relative to the size of the company) against Harvester than against Deere, although both firms were clearly attempting to get dealers to sell as much of their line and as little competing equipment as possible. Harvester's rather more heavy-handed approach may have been due in part to its having to rely on only one dealer per town—however densely farmed the area. This competitive handicap was the result of a consent decree of 1918 when the firm was widely believed to be attempting to monopolize the industry.[1] The difference in product line and dealer strength undoubtedly account for much of Harvester's vastly great vulnerability to competitive thrusts from smaller market participants. From a 60 percent share in 1929, Harvester's share dropped almost continuously afterward.[2] Allis-Chalmers' new designs in the thirties appear to have shifted many of International Harvester's customers but almost none of Deere's, and during and after the war, the Ford-Ferguson blitz drew much business from both International Harvester and Allis-Chalmers (which specialized in the most directly competing small, maneuverable tractors), but left Deere relatively unaffected.

Post-War Distribution and Product Line Developments

All of the major firms (including Ford's and Ferguson's distributors) offered some dealer and retail finance after the end of the post-war boom which had caused a virtual halt to this traditional industry function. There appear to have been considerable differences, however, in the extent of company re-entry into finance at both levels. Unfortunately, it is impossible to document such variation accurately because none of the firms has an historical record of receivables by product line, and some even fail to distinguish between wholesale and retail credit. It is clear, however, that in 1957 Deere permanently increased the extent of credit provision in the industry.[3] A dip in farm income plus a 19-week strike cut Deere's 1956 sales by 8 percent, and the firm decided to use its strong working-capital positon to fight back. Beginning in 1957, Deere gave its dealers authority to accept as full payment for all machinery any farmers' notes which could be regarded as "sound." Deere's retail receivables shot from $38 million at the beginning of 1956 to $132 million at the end of 1958, the year it overtook International Harvester as the leading farm machinery seller in North America. As late as 1960, Harvester followed a different course. While apparently providing adequate dealer finance, the latter's policy was "to encourage its dealers to finance retail time sales with banks and other finance companies. . . ."[4] As a result, its retail finance increased only from $8 to $30 million between 1956 and 1959. It appears that the smaller full-line firms were already following Deere's lead in the retail credit field, however, and International Harvester's policy turned around sharply in the early sixties. Subsequently, the terms of retail sales became rather uniform among the major manufacturers, and retail credit largely ceased to be a competitive instrument.[5]

Why did the industry become so heavily involved in the financing of retail sales, and, if such financing became the *sine qua non* of effective marketing, why weren't the funds necessary for its effective functioning included in the capital requirements in Chapter Three? W. G. Phillips, an authority on the North American industry concludes:

> The reason the industry assumed such a heavy credit burden seems to be two-fold. First, the lending policies of most banks have proved to be not readily adaptable to the particular needs of large-scale farm machinery credit, leaving the industry little practical alternative to assuming the credit load itself. Secondly, the operation itself is by no means unprofitable by nature. . . .[6]

What the former part of the explanation ignores, however, is the possibility that a finance company specifically organized to serve the farm-machinery market be a serviceable alternative to either bank or company finance. Indeed, White Motor during the sixties relied on such arrangements with independent financing firms (of which there were several operating at the national level).[7] Deere at the time of its credit blitz, however, was almost exclusively a seller of farm machinery and may have perceived a low opportunity cost of funds in the shrunken but still profitable market.[8] This is consonant with the other part of Phillips' explanation. After analyzing data for the sixties, Martinusen and Barry conclude that "it has not been possible to state whether the credit-granting operations are competitive diversification on the part of the firms or a selling aid for traditional product lines."[9] It seems reasonable (although conservative) to conclude that in-house retail credit facilities, while perhaps desirable, are not really essential, and therefore do not deserve to be included in capital requirements estimated for the late sixties in Chapter Three.[10]

An extremely important North American development during the more recent years of the period was the evolving strategy of Ford. After the split with Ferguson, the company continued to concentrate its efforts on small tractors, implements for use therewith, and slightly modified non-farm units (which led the industry); but in 1964, as part of renewed corporate attention to tractor selling after a substantial decline in market share since the mid-fifties, Ford stopped selling through its 24 independent U.S. distributors while retaining nearly all of its retail outlets. The reasons given were that economies of scale could be gained by dealing through fewer wholesale points (but, of course, this could have been accomplished by changing the territories of existing distributors and dropping others) and that the move would have the effect of "bringing us closer . . . to serve our customers more effectively" which a trade journal report interpreted as meaning that Ford would have "a closer, more exclusive, relation with its dealers."[11] The planned extension of Ford's product line which was to include more implements and larger tractors could therefore be introduced through dealerships without any difficulties from the competing equipment which the independent wholesalers had handled, and the attention of dealers to Ford products would be subject to direct monitoring. After adding the imported Claas combine in 1966, considerably broadening its implement offerings, and introducing high powered tractors in 1969, Ford had arguably become full-line, although its implement range was still somewhat less complete than that of most of its competitors.

During the period of product-line expansion Ford also attempted to recruit dealers in the most lucrative part of the U.S. market, the Midwest, where it was traditionally weak. This recruitment usually took the form of winning established dealers away from other brands, and Ford's ability to do this had clearly become much greater by the late sixties when it was able to substitute one virtually complete line for another.

The other firm which made conspicuous attempts to upgrade its distribution system in North America, especially in the U.S. Midwest, was Massey-Ferguson. For several years after the 1953 merger, Ferguson and Massey-Harris dealership systems were run independently. Massey-Harris's "lugging" tractors, however, were clearly obsolescent, and subsequent company research and development ultimately undermined the "two-line" policy. It was simply not feasible to produce two tractor lines of comparable quality and range with non-trivial differences. Ferguson's independent distributors were therefore bought out in 1957.[12] Throughout the late fifties and sixties, Massey-Ferguson attempted to increase its dealership strength, and in promising areas where suitable dealers could not be found, the firm (with proclaimed reluctance) opened company stores.[13]

Price Policy and Structural Change

Chapter Ten will document the enormous price differences that have existed between similar machines sold in Europe, particularly in the U.K., and North America. It will also become clear that although production cost per horsepower declined sharply with increasing machine size, price per horsepower in North America in the late sixties actually rose over certain mid-ranges of power and, even on the most powerful units, allowed a vastly higher profit per horsepower than on the smaller machines. This raises the question: Upon what basis are North American tractor prices determined?

International Harvester appears to have been the usual North American price leader, at least until the late fifties, although in the pre-war period leadership was sometimes shared with Deere.[14] The price changes are usually announced during the weak selling season around November 1, but it should be stressed that, unlike the situation in the automobile industry, these annual price revisions are more often than not simply the repricing of an essentially unaltered line (see Chapter Nine). If new models are introduced, however, they typically accompany the price adjustments. An investigation of the timing of price announcements for tractors made by the RCFM (prices are announced simultaneously for the U.S. and Canada) for the period of 1963 through 1967 found Deere announcing first (and without subsequent revision) for every year but 1967 (when International Harvester was again first).[15] It is probably not a coincidence that during the period for which no information on price leadership is available, Deere overtook International Harvester as the largest seller of farm equipment in the U.S. Tractor supremacy is harder to document, but this too was Deere's by at least the late sixties.[16]

In the previous chapter it was noted that the prices of the major firms, taking horsepower and special equipment into account, have differed but slightly in North America (as elsewhere). It is further true that no brand has been conspicuous by being cheapest in all power classes. Nevertheless, in a careful study of Canadian prices for similar models in 1967, David Schwartzman [17] found that one of the small American full-line firms, Allis-Chalmers, Case, or White, was cheapest in the 60 to 75, 75 to 90, 90 to 100, and 100 to 115 horsepower classes, which together accounted for 56 percent of the total unit demand in that year. In addition, as was explained in Chapter Seven, British Leyland and David Brown established themselves with increasing shares in the low and medium horsepower ranges apparently in large part as a result of a consistent and pronounced low price appeal.

Table 8-1 shows the fragmented character of the Canadian market in 1967, partly as a result of the price differences just outlined. It is evident that the performance of the companies was very uneven, and that at least one of the smaller firms had succeeded in becoming one of the four leading sellers in nearly every power category. In the 40 to 59 horsepower category, David Brown machines were in the top group, and in the three highest horsepower

Table 8-1. Sales Concentration Ratios for Wheeled Tractors, Canada, by Size Class, 1967

	Percentage of Total Unit Sales	
	Four Largest Firms	Eight Largest Firms
Under 40 HP	86.7 Deere, Ford IH, M—F	100.0 A—C, Case D. Brown, White
40 to 59 HP	58.6 D. Brown, Deere Ford, M—F	85.9 A—C, Case, IH White
60 to 69 HP	62.1 Case, Deere IH, M—F	94.9 C.C.I.L., D. Brown British Leyland, White
70 to 99 HP	84.4 Case, Deere IH, White	100.0 A—C, M—F, Ford
100 HP and over	81.8 Deere, IH Versatile, White	100.0 A—C, Case M—F

Note: White includes all tractors sold by Cockshutt and Minneapolis-Moline (subsidiaries of White Motor Company) and Oliver tractors sold by Coopérative Fédérée de Québec. Firms in each group are listed alphabetically.
Source: Barber, Prices, p. 10.

categories, the tractors of the smaller U.S. full-line firms were making a strong showing. Case appears as a leader twice, as does White. Tiny Versatile was able to attain top four status in the highest power category which at the time accounted for about 10 percent of industry sales.

Although 1967 is the only year for which the market positions of the firms are identified, time series material collected by the RCFM indicates that the combined shares of Deere, International Harvester, and Massey-Ferguson in the value of total Canadian tractor sales dropped considerably over the period of 1957 to 1967, from 62 percent to 51 percent, and that of the four leading firms dropped 5 percent from 72 to 67 percent, while the six-firm ratio rose one point from 86 to 87 percent.[18] The six largest firms, at least in the sixties, were undoubtedly Deere, International Harvester, Massey-Ferguson, Ford, Case, and White. Although the share material is available only for Canada, the broad outlines of the pattern undoubtedly hold for the U.S. as well, and one is left with the problem of explaining why the largest firms allowed their collective market share to be eaten away. To provide an explanation, David Schwartz-man and Clarence Barber develop a numerical model which illustrates the market impact of equal percentage price cuts by firms which are either "large" or "fringe"[19] They argue that the smaller firms might not take away enough market share to induce immediate retaliation and that it might be consistent with the well-being of the industry giants to allow the smaller firms to grow and prosper to a certain extent, because the speed at which profits are competed away is the critical consideration. Such analysis is familiar enough in the theory of industrial organization;[20] what is less clear is how the theory fits the U.S. and Canadian tractor markets—Schwartzman and Barber's model makes no distinction between foreign and domestic firms. It may well have been profitable to allow some attrition to David Brown and BLMC in the low and medium horsepower ranges rather than fighting them.

But what of Case, the White companies, and Allis-Chalmers? These firms, as the RCFM establishes, had disproportionately high sales in the most profitable part of the market. Together they accounted for some 55,000 tractors or 20 percent of unit output in 1966 and a still higher fraction in value. One would guess that a few years at most of lower margins on large tractors, and perhaps on combines for good measure, would have driven out the smaller full-line firms. From 1957 through the last year of independent operation for which data are available, the companies sold to White had very modest average net income as a percentage of equity: Oliver, 2.5 percent; Cockshutt, 3.7 percent; and Minneapolis-Moline, almost no profit at all (.04 percent). The firms were acquired in 1960, 1962, and 1963, respectively. J. I. Case came under the control of the conglomerate Tenneco in the late sixties and lost its independent existence completely in 1970. Nevertheless, over the 1957 to 1967 period for which the gain in market share for the smaller firms was recorded, it was the only small company which produced and sold a full-line of equipment in North America and had over half of its total sales in farm machinery. Case's average return on equity over the period was minus 1.1 percent,[21] and Allis-Chalmers (under 30 percent of the sales of which were in farm machinery) was in deep financial trouble over the 1957 to 1967 period. The farm machinery operations of all of these firms would appear to be in a position of "short-run vulnerability" through price-cutting by competitors which Martin Shubik ascribed to American Motors and Studebaker-Packard in the U.S. automobile industry in the mid-fifties.[22]

Two factors may account for the observed restraint of the larger companies. First, one of the leaders, International Harvester, had experienced considerable attention from the U.S. government, admittedly most of it before the war. Nevertheless, company policy as stated in the mid-fifties was "to bend over backward in its interpretation of the law,"[23] and predatory intent could possibly be attributed to participation in such an activity by any of the large firms.

A second possible factor is that the larger firms may not have wanted to force sales of the smaller ones, since that could have facilitated international entry. White was able to buy Minneapolis-Moline at 60 percent of book value and Oliver at 80 percent of book;[24] as a result, the new operations were claimed to be earning reasonable profits for a few years thereafter. It was well known after the late fifties that several European firms were interested in selling more tractors and other farm machinery in North America; British Leyland, David Brown, and Deutz did so on a small scale. Yet both Renault (the machines of which were ineffectively marketed in Quebec by a cooperative, see p. 135), and Fiat claimed not to have made a serious effort in North America because of the formidable problem of establishing a distribution network.[25] Further, Fiat helped organize a loan for Allis-Chalmers in 1969, and was associated with the U.S. firm in Europe in a joint venture to produce and distribute construction equipment.[26] The larger firms may not have wanted to force the smaller U.S. operations to the block, in part because of the temptation offered to foreign firms, which, of course, would undoubtedly be far more interested in the established names and distribution systems than in production facilities. The authorities would probably look askance at the acquisition of a "failing firm" by a strong domestic competitor, while they might well not do so if the acquiring firm were foreign.

This suggests a general proposition: market situations in which price competition would cause smaller rivals ultimately to fail in a closed economy may be avoided when outsiders with advantages in production volume or absolute costs could use the acquired firm as a means of overcoming the product differentiation disadvantage from which they would otherwise suffer.

Product Strategy of the Small Full-Line Firms

The annual reports of Case and the White companies from the early sixties onwards indicate that concentration on larger tractors was central to their strategy. Furthermore, Chapter Ten will show that the smaller U.S. full-line firms would have actually been losing money on sales of lower h.p. units by the late sixties if operating from a new plant. It is not surprising therefore that they thinned out or abandoned smaller models in the sixties, offered them "packed" with non-optional features, failed to make even the most minor revision in them or, most significantly, imported small foreign units to be sold as part of their product line. Case's 1958 purchase of Vierzon, noted in the previous chapter, was almost wholly to give it a new lease on life in the small horsepower diesel market. It was subsequently discovered that the expense necessary to improve the Vierzon design for the North American market would be greater than that of continued production of small diesel machines in the U.S., and the plan was abandoned. This was followed by Cockshutt's purchase of Fiat machines for sale in the U.S. and Canada in 1960. Oliver sold one David Brown model from 1960 to 1966, and White Motor, after having gained control of both Cockshutt and Oliver, began marketing slightly modified Fiats in 1965, a practice which continued into the early seventies. Allis-Chalmers completed the small firm practice in 1969 by selling a slightly modified Renault machine as its smallest model.[27]

It is an interesting confirmation of the "image" speculation of Chapter Four that, despite the fact that Fiat was the fifth largest manufacturer of wheeled tractors in the world by the early sixties, its manufacture of two small White tractors for the North American market was not only unpublicized but apparently deliberately suppressed. It is true that the suppression could have been partly to avoid drawing purchaser attention to the issue of why White chose to do foreign sourcing (its probable inability to make the machine profitably itself), a desire to retain maximum flexibility in sourcing, and a wish to render more difficult the discovery of international price differences. The company's professed concern about farmers' reaction to a tractor which needed slight modification because the grill was too "Italian looking,"[28] however, might have stemmed from a fear that any kind of Italian identification should be avoided if possible.

Industry Consolidation

Competitive pressures also induced two important industry consolidations. By far the most important was the merger of Massey-Harris and Ferguson in 1953, a move which was examined but not opposed by the Federal Trade Commission, apparently because the firms' product lines were largely complementary rather than competitive. Furguson's only product was a tractor with implements for attachment thereto, while, although Massey-Harris offered tractors, virtually all of its strength lay in harvesting equipment, particularly in its self-propelled combines which accounted for over half of all U.S. sales in 1948, 1950, and 1951.[29] Without a broad and modern tractor line, however, Massey-Harris was losing ground rapidly in the U.S. in the early fifties. Ferguson's weak distribution system by comparison with Ford's was causing him to take a drubbing as well, and the "failing firm" view was probably important in official circles.[30] At the time of the merger Ferguson had a 5 percent share of the tractor market and Massey-Harris just over 3 percent, although Massey-Harris's share in farm machinery was several points higher.[31]

The other merger was actually a takeover of three separate companies—Oliver, Cockshutt, and Minneapolis-Moline—by White Motor between 1960 and 1963. All three were purchased at bargain prices which presumably reflected a recognition that the part of the market they could profitably serve in the long run was understood to be limited. White apparently believed that its expertise in wheeled vehicle production and distribution could bring the operations out of the red, although their combined volume was less than 20,000 units. Despite the discontinuation of Canadian manufacture which came after the White acquisition, machines sold in the U.S. as "Oliver" were sold in Canada as "Cockshutt" (except in Québec)—an example of recognition of the kind of brand loyalty which the a priori argument of Chapter Four would lead one to expect. Separate manufacturing facilities and dealerships for Minneapolis-Moline and Oliver (Cockshutt) were retained until 1972 when it was reported that the White farm machinery dealerships were being consolidated and Minneapolis-Moline production facilities closed, thus spelling the end of Minneapolis-Moline in fact if not in name.[32] If the White companies are treated as a single entity, in 1970 there were no producers of a complete range of wheeled tractors in North America operating at a scale substantially lower than 15,000 units per year, while the largest North American operations, those of Deere and International Harvester, produced only about four times that amount.

THE UNITED KINGDOM

Developments in Distribution

The manufacturer acting as his own wholesaler became the prevailing practice in Britain, although this was retarded by Ferguson's early importance. Harry Ferguson had built up an effective distribution system based, as in North America, on independent distributors, but as the Massey-Ferguson tractor share began to decline in the late fifties, the company began to sense that the system of very large distributors was causing difficulties. Such dealers had an undesirable independence of policy and a leverage over matters of pricing and even product design. Their elaborate network of subdealers, some selling only tractors or only combines, had reduced information and control and had kept the manufacturer too far away from the final purchaser in both sales and after-sales service. Massey-Ferguson's management took action to divest itself of the system, and in doing so lost a distributor handling over 10 percent of its British sales to another firm (David Brown) in 1960.[33] The immediately following period was the low water mark for the firm in terms of market share, which grew in the next few years, presumably in part as a result of the distribution system change.

Ford, the other giant in the U.K., appears to have handled its own wholesaling throughout, but it has relied on many more dealers than Massey-Ferguson for about the same sales volume. This was economical for Ford because many of those outlets also sold cars and trucks.[34]

In the pre-war and early post-war period, there was apparently rather little exclusive dealing in the U.K., and customers were usually able to get many makes of equipment through one dealer. By the late fifties, however, the major companies, drawing from North American experience, moved toward exclusive dealing, with its potential for "encouragement" to sell the complete offerings of the manufacturer. As in North America, the typical U.K. dealer in the sixties usually handled only one make of tractor and the associated line of implements, although multiple franchises were not uncommon and usually involved the smaller companies.

In the early and mid-fifties exclusivity was attacked by the Agricultural Machinery and Tractor Dealers' Association (AMTDA) as "un-British," but, as time wore on, increasing acceptance was achieved. This acceptance may have developed in part because of its potential for contributing to "orderly trading" through territorial exclusivity in the face of important legal changes affecting the tractor trade. In the early fifties, the AMTDA had hoped to revive a pre-war scheme with the Agricultural Engineers Association (the manufacturers' group) to collect a part of the dealer discount to be paid out only when all conditions of AMTDA membership had been satisfied. The most important of these conditions would have been the avoidance of excessive trade-in allowance which was the easiest way to circumvent the prevailing regime of resale price maintenance. A sufficient explanation, though not necessarily the only one, as to why this scheme never got off the ground was the discovery that the kind of activity proposed would probably appear to be restraint of trade of the kind attacked by the Restrictive Trade Practices Act of 1956.[35] In 1964, resale price maintenance itself was made illegal, although territorial exclusivity continued unchallenged.

In the U.K. manufacturers serve an area of roughly the same size as that of Illinois and Indiana with an even smaller area of agricultural land. Nevertheless, one-quarter of the annual U.S. unit volume was sold in the United Kingdom in the late sixties. Regional wholesale branches are unnecessary, and machines usually move directly from the factory to a "main" dealer. The tractors are sold by these franchised dealers (with spare parts service) and by their appointed sub-dealers (if any) who work on a smaller margin and who are often little more than "bird-dogs" for prospective customers.[36] The major manufacturers have never provided substantial credit to either dealers or final purchasers.

Behavior and Structural Change

The impact of entry and subsequent internal competition on the British market was to turn a virtual monopoly (Ford) into a duopoly (Ford and Massey-Ferguson) with at least two healthy fringe participants: David Brown and British Leyland. There were no mergers except for Massey-Harris and Ferguson, and the exits were of those small firms foolhardy enough to try to compete selling ordinary wheeled tractors; virtually all had expired even before they developed Ferguson-type hydraulics. Minneapolis-Moline's venture was practically still-born, and Allis-Chalmers, maintaining a very thin product line and a miniscule market share throughout, finally withdrew in 1970.

International Harvester, which did not develop Ferguson-type machines until the late fifties, enjoyed very limited success in the U.K., but it still offered British-produced models at not unusual prices in the early seventies. Like most of the British industry, it exported more than 70 percent of its output in most years.[37] For Massey-Ferguson and Ford, this meant British production of many tractor components, including engines, well in excess of 100,000 units per year.

Britain's advantage as an area of low absolute cost is illustrated by the experience of David Brown and British Leyland. While David Brown had an output of only about 10,000 units in 1960, this was up to 24,000 by 1967 and around 30,000 by 1970. British Leyland averaged a yearly output of only 7,500 units between 1948 and 1964, although its 1963 output had risen to over

13,000 units. An entirely new plant was completed in 1964, and the firm's output climbed to about 16,000 units in 1969. The number of tractors which stayed in the home market is not available by year, but, although both firms usually claimed over 80 percent of output exported, company statements also imply that the home market share of both firms went up several points in the sixties—each still had well under 10 percent of the market in 1970.[38]

The domestic sales of the two firms are not well enough recorded to make possible any close examination of the relation between their price policies and market shares. What can be observed is that while price lists as late as 1960 reveal no significant differences between the machines of these firms and those of Ford and Massey-Ferguson on a price per h.p. basis, as the sixties wore on, differences became much more noticeable, and in 1967 prices were between 5 and 10 percent lower on most David Brown and British Leyland Models.[39]

The success of David Brown and British Leyland is attributable to two factors. In the home market Ford and Massey-Ferguson presumably saw little threat from these firms, with their limited range of models and exiguous ancillary offerings, and so allowed them to sell at a noticeable price advantage; at the same time, in many foreign markets the small U.K. firms simply exploited the "umbrella" of high prices provided by the domestic price leaders (see Chapter Ten).

As was the case in North America, the only tiny firms with staying power were those which over the years served very specialized parts of the market, most notably those offering relatively powerful four-wheel drive units. Like their North American counterparts, Versatile and Wagner, Doe, County, Roadless, and Muir-Hill seem to have survived by selling tractors offering a maximum of specialized design and a minimum of value-added. (Doe, County, and Roadless were all modifications of Ford units.)

Although producers from Eastern Europe used spectacularly low prices with some success after the mid-sixties, there was no evidence as late as 1970 that they had noticeably affected the main participants' conduct.

FRANCE

Distribution and Product Line Developments

The North American distribution system presented a simple pattern of direct major firm wholesaling to the dealer level with Ford and Ferguson as historical exceptions. The British case added some complexity with the greater historical importance of independent distributors on the one hand and of the dealer cum sub-dealer arrangements on the other. In other markets, the pattern has been even more complex, and the only unifying theme seems to be the secular growth towards the North American system of direct relations between company and dealer.

The many small firms which served the French market during the first decade after World War II sold their machines through any channel that was available. Erstwhile horsetraders seem to have played an important role, but, as the importance of after-sales service became apparent and many of the smaller manufacturers fell by the wayside, there was a considerable consolidation in retail distribution. The nominal dealer force may have been reduced by as much as three-quarters between the late fifties and 1970, while tractor sales changed very little. In addition, however, Renault, the second largest seller, was channeling a large part of its output in the mid-sixties either through factory branches or sub-dealerships responsible directly to them. In the late sixties it also attempted to boost its sagging market position by putting company stores (*succisales*) into areas which couldn't support an independent dealer (and presumably where sub-dealers had proved to be unsatisfactory) and bolstering the sales of these stores with the full range of Renault trucks and cars. (How the scheme fit in with previous car and truck distribution patterns was not reported.) At the same time Renault was only beginning to establish some strong independent dealerships of the kind found in North America and Britain. In fact, while Renault boasted in 1968 that it had an outlet of some kind within six miles of every French farm, Massey-Ferguson, relying solely on 230 large independent dealers, was outselling Renault at the time and had been almost since the beginning of the period. There has been a secular increase in exclusivity over the years, although only Massey-Ferguson claimed that all of its dealers were exclusive by the mid-sixties.[40]

Trade sources indicate that Ford's distribution was considerably "up-graded"[41] in France in the early sixties as its share increased, but this could have occurred *pari passu* with increased demand rather than playing an important role in generating it. It may be no coincidence that by the late sixties when Ford's models covered perhaps 90 percent of the market, its share leveled off. Especially in the last years Ford probably suffered some competitive disadvantage from offering only a partial machinery line to a national dealer body which was becoming increasingly exclusively franchised with full lines—the lack of a combine was particularly damaging.

Although dealer credit has not been an important feature of French distribution, Deere's strategy in the sixties was to work toward a stronger distribution system by following North American practice. A spokesman in 1966 stated, "We offer the same terms we do in the United States. . . . our dealers buy on open account and stock until sold. This lets them keep inventory for a season or a season and a half."[42] The spokesman claimed that some of the other companies were beginning to do the same thing.

The most unusual feature of French distribution by contrast with the systems discussed so far is the role of cooperatives, despite their complete avoidance by Massey-Ferguson. Renault sold 30 percent of its output through cooperative channels in 1970;[43] the significance of cooperative distribution will be evaluated at the end of this chapter.

Renault was the only French-owned firm to survive the competition of the late fifties and early sixties, and competitive pressure induced an important change in its product line as well as in distribution practices. Of Renault's three largest competitors in the sixties, two, Massey-Ferguson and International Harvester, were full-line firms, and, while Ford's increase in the early sixties was very dramatic, the most conspicuous steady rise in the latter part of the decade was that of the full-line firm, Deere-Lanz. Furthermore, Deutz, also an impressive gainer in the market, had acquired 32 percent of the German combine firm, Fahr, by the mid-sixties. As an expressed response to the growth of full-line dealing by competing firms, and a concomitant growth in exclusivity, Renault announced in 1968 that it had entered into a joint marketing arrangement with several private firms, the most important of which was the only French-owned combine manufacturer, Braud. The result of the complex arrangement was to offer "a worthy rival . . . to the great American organizations."[44]

Structural Change

The share of the four leaders in France, Massey-Ferguson, Renault, International Harvester and Fiat, grew from 61 percent in 1952 to 77 percent in 1959, but receded again to 64 percent in 1965 and 62 percent in 1970. This decline was the result of the increased strength of importing firms—Ford, Deutz, Deere-Lanz, and David Brown—which each held only between 1 and 2 percent of the market in 1959, but the combined share of which had grown to 23 percent by 1970. They pushed Vendeuvre and Vierzon out of the top eight and into oblivion by the mid-sixties, while smaller domestic firms producing standard wheeled tractors met their demise in the fifties. Although as late as 1970 there was a handful of extremely small firms specializing mainly in high clearance machines, the largest had but .7 percent of the market, which implied production of only a few hundred units, even allowing for some exports.[45] Some of these small producers were descended from earlier producers of standard tractors. Even Renault, the largest producer relying entirely on French components, had a yearly output of under 20,000. There is no doubt that constant price pressure on the market from foreign manufacturers, particularly Ford, played an important role in consolidating the market and therefore changing its structure, although the result was a considerable lessening of top-level concentration.

GERMANY

Distribution and Product Line Developments

By almost any standard the German distribution system, as late as the early seventies, was the most varied of any of the markets in the study. Even defining a retail outlet was difficult because in the mid-sixties, in addition to the regular dealer force, many of the country's 20,000 blacksmiths also sold farm equipment, including tractors. In the fifties it was apparently common in at least part of the industry to have different classes of dealers, each with different terms, or even dealer-by-dealer bargaining.[46] In a statement in 1966, William Hewitt of Deere reviewed his firm's experience over the preceeding nine years. "In Germany . . . dealers generally carried a number of lines, and each dealer negotiated with the factory on the price he should pay. We have expended a lot of effort to try to create a more logical dealer system in the market."[47]]

In Germany in the sixties it was more common than in France for the major sellers to pursue multiple modes of distribution. Massey-Ferguson, all of the French outlets of which were independent dealerships, was rapidly expanding its German participation in 1963 through a complex network of exclusive dealerships, retail cooperatives, independent distributors (which were often both retail and wholesale outlets), and hundreds of "casual" dealers who sold only perhaps one or two tractors a year. Of the company's total German outlets, only 20 percent were claimed to be exclusive in 1965.[48]

Both exclusive dealing and "full-line" selling—the English term is used—while considered by many to be inevitable were considerably less well developed than in France by the early seventies. The company with the strongest dealers—"virtually all exclusive"[49]—was first-ranked Deutz, with second-place International Harvester judged to have the next largest percentage of exclusive dealers but to be a long way behind the leader.

Deere's retail outlets, inherited from Lanz, were judged to be wholly inadequate by U.S. management. The Lanz dealerships usually handled a number of lines, but, by 1966, most of Deere's dealers had become exclusive.[50]

The usual practice in Germany has been that of a dealer financing his own inventory by cash or through finance companies, but by the late sixties some manufacturers had begun to use dealer credit as a competitive device. Some firms offered credit to dealers at interest rates several percentage points below the market for only a small down payment, while, as in France, Deere made the most lavish credit provision, matching or bettering American floor-planning terms. Small manufacturers saw the increased need for working capital implied by the new marketing practices as a threat to their very existence, although those which failed in the following years probably would have done so anyway.[51]

Although meaningful national figures on the number and size of cooperatives in German distribution are not avilable, they do play a significant role. Cooperatives serve not only as retail outlets but also as distributors. For example, the Württemberg Central Cooperative in the late sixties was a regional distributor for both the domestic firm, Fendt, and Fiat.[52]

Although most tractor-makers offered or endorsed some implements after the mid-fifties, an important factor retarding the growth of North American-style distribution was undoubtedly the fact that Claas, the world's largest combine maker in the sixties, and with 60 percent of the German market, did not offer a tractor and found the prevailing distribution system satisfactory. Nevertheless, the full-line firms troubled the leading domestic seller just as they did in France. Not only was Deutz's chief competitor, International Harvester, a full-line firm, but the most rapidly growing firm in Germany in the sixties was another full-line seller, Massey-Ferguson. In response to the growing strength of firms with broader product lines, Deutz began acquiring stock in Fahr, the country's second largest combine manufacturer, and acquired two important implement firms. Almost simultaneously with Renault's announcement of the link with Braud in 1968, Deutz announced that it had acquired majority control of Fahr. Some industry opinion held that the move narrowly averted a takeover by International Harvester which was seeking additional combine strength in Europe (it already had one plant in France).[53]

Ford's arrangement with Claas to market its combine was confined to the North American market, and no such arrangement was reached in Europe, although some have predicted it.[54] Perhaps increasingly for lack of a full-line, Ford's performance in Germany after its spectacular price moves discussed in the previous chapter worsened noticeably; its high of 5 percent of the market in 1965 had fallen to 2.7 percent in 1970.

Behavior and Structural Change

In 1956, the first year for which seller shares are available, there were an estimated 85 firms in Germany producing 140 tractor models and accounting for a production of 135,000. Seventeen German firms (including International Harvester's German facility) claimed over 1 percent of the domestic market, as did Massey-Ferguson importing from France.[55] 1955 and 1956 were the peak output years for the German industry, and even then no more than three firms approached an output of 20,000 units, a level which was probably never reached again up to 1970.

After the mid-fifties, a multitude of small firms disappeared, and it is impossible to discern the relative importance of the increasing appeal of certain domestic firms (including International Harvester) as against price pressure from firms importing into the country, particularly Massey-Ferguson, in bringing this about. Deutz, however, is reported to have been the usual price leader [56] and seems to have been reluctant to attempt to expand its domestic share through price competition, perhaps fearing that some of what were, for the German market, medium-sized firms would be driven into the hands of large and dangerous foreign rivals, able and willing to make more attractive offers than Deutz itself. This may help to explain how, as late as 1969, several domestic firms selling ordinary farm tractors at very low volumes were surviving. These included Hanomag and Eicher, each with annual production of probably less than 5,000 units. There were even smaller domestic firms selling to specialized parts of the market as in France, but presumably none of them was making its own engines or other major components. Third-place Massey-Ferguson bought a 30 percent share in eighth-place Eicher in 1970, and that year was also the last for sixth-place Hanomag which simply ceased production after a brief attempt to increase market penetration by cooperating in distribution with Daimler-Benz's "Unimog," general purpose vehicle.[57] These events left Deutz (20,000 units) and Fendt (10,000 units) as the only wholly German-owned producers of a standard range of agricultural tractors.

It is ironic that the North American firm most market-oriented and so successful at home should have done so poorly in the home market of its only foreign acquisition. Although Deere-Lanz achieved a 400 percent increase in neighboring France, which gave it a share there similar to that held in Germany, in the latter market it could do no more than hang on to an unprofitable 5 to 6 percent market share. In 1971, with a production of fewer than 20,000 units and after more than a decade of continuous losses (estimated at $59 million by the end of 1967), it was announced that Deere's European operations were to be combined with those of Fiat in both agricultural and construction equipment on a 50–50 basis. A Deere spokesman at the time noted that such an arrangement was subject to prior approval by the authorities. The proposed venture was abandoned by the end of 1971 with a joint communique noting that the two firms simply could not agree upon terms; there is no record of governmental difficulties.[58] An interpretation of these negotiations will be given in Chapter Eleven.

Increasing concentration in Germany and decreasing top-level concentration in France left those two countries with very similar concentration profiles by 1970.

ITALY

Developments in Distribution

In the early fifties the distribution of capital goods for agriculture including tractors was largely controlled by *Federconsorzi,* an organization for the co-ordination of provincial cooperatives. Although it claimed at one point to purchase 80 percent of the modern material inputs used in agriculture, and Fiat, the industry leader, continued to market through cooperative channels throughout, some important firms did not participate; an example is Landini, the third largest firm in the fifties.[59]

Exclusive dealing was claimed in a trade journal article in 1967 to have become as important as in Britain, but full-line selling has not been as important an issue in Italy as elsewhere, for, although many tractor makers do sell some implements, the self-propelled combine is not widely used in Italy. Only 1,600 were produced in Italy in 1965 and sales in that year were 2,312, or 1 combine for every 14 tractors. No Italian tractor-maker even offered a combine with its name on it. Another characteristic distinguishing Italian distribution from that in most other markets (except the U.K.) is the apparent absence of manufacturer credit to retailers.[60]

All importers except Ford and Massey-Ferguson worked through independent national distributors, and the very low level of internal prices in Italy during the sixties (see Chapter Ten) probably explains why it was the only country in the study in which David Brown did not attempt its low price strategy; the somewhat less consistent low price firm, British Leyland, also did not offer low prices in Italy. Both firms quoted prices higher than those of several other importers and much higher than those of Fiat.[61]

Increasing Concentration and Specialization

The increasing concentration observed in the Italian market appears to have been almost wholly an internally generated phenomenon. During the sixties, the role of imports in the Italian market was diminishing and foreign suppliers were suffering from the same kind of price competition as the small domestic firms. The top eleven wheeled tractor sellers in 1964—four domestic firms, one foreign firm with domestic production, and six importers—accounted for all but 10 percent of units sold; those same firms in 1956 accounted for only 70 percent of sales, and virtually all of the other participating firms had less than 3 percent of the market, or annual sales of a few hundred units or less.[62]

Despite their decline, small makers still had a substantial part of the market as late as the mid-sixties. A close look at the actual models offered, however, renders the difference from other markets considerably less significant. Italy's difficult terrain produced a unique market for small (but not necessarily low-powered) four-wheel drive machines of a very different overall configuration from that of a standard agricultural tractor; most designs were based on a pivoted frame for steering. While there were a dozen or more makers of such machines in 1964, all engines were purchased from the outside, and, if other agricultural machines were sold by these firms, they were almost invariably moto-tillers rather than standard farm tractors.[63]

A merger of perhaps more apparent than real importance was the official linking of first-place Fiat with fourth-place OM in 1959. It is highly possible that much of OM's stock had been owned previously by Fiat because it is probably not a coincidence that the firms had never produced competitive models, and Fiat's power range had gaps in it corresponding to the machines sold by OM. This pattern continued after the formal combination of the firms.

In 1970 there were only five firms in Italy making agricultural tractors of the kind almost universally used elsewhere: Fiat-OM (40,000 units or more), Same (almost certainly not more than 10,000 units), Landini (perhaps 5,000), Lamborghini (3,000) and Carraro (fewer than 2,000). What is truly extraordinary is the ability of the two tiny independent firms to stay in the race, especially because both produced their own engines.[64]

AUSTRALIA

Developments in Distribution

The largest sellers in Australia—Massey-Ferguson, Ford and International Harvester—operate their own state branch wholesaling and also employ exclusive franchises. As elsewhere, Massey-Ferguson did away with the separate Massey-Harris and Ferguson distribution systems and established a unified system along North American lines in the late fifties. When the firms with only North American sources of supply re-entered the market after dollar restrictions were lifted in 1957, only Deere moved to completely controlled wholesaling in the small and geographically dispersed market. In 1970, Allis-Chalmers and Case directly controlled some of their distribution to the dealer level while employing independent distributors in some states as well, but the White companies, much smaller in Australia than the other minor U.S. full-line firms because of the lack of construction sales to augment their agricultural offerings, employed only one independent distributor for the whole country. Fiat operates through independent distributors in every state, and cooperatives serve as state distributors for several lines, including the smaller U.S. full-line firms, apparently on standard wholesale terms.[65]

Rural areas in Australia, as in North America, are apparently not blessed with a surfeit of promising potential farm equipment dealers, thus confirming our prior suspicions. By the early sixties, "With the exception of one big company, all firms admitted that they have difficulties in finding good dealers."[66] The big company may well have been International Harvester, because Massey-Ferguson was then considering setting up company stores in areas in which it could not find suitable dealers. More than in other markets, because of the extreme sparseness of the population in many areas, dealers often sell not only both farm and construction equipment but cars and trucks as well.[67]

Exclusive dealing, exclusive territories, and resale price maintenance were all legal under Australian law until the sixties. The first two became contestable under the Trade Practices Act of 1965, and the third may become contestable as interpretations of the Act are developed.[68]

As late as the early sixties, only very short credit terms to dealers were offered, but, as on the Continent, this was followed by a period of increasingly liberal credit. By 1970, however, the Australian system was still quite different from that of North America because the dealer faced a choice of discounts depending on payment terms. [69]

Industry Behavior, Government Subsidy, and Structural Change

Although nothing is known of price leadership in Australia, it is of interest that the strategy of the sole domestic competitor, Chamberlain, appears to have been much like that of the small North American full-line firms in their home markets—avoiding serious conpetition in the low horsepower classes and aiming its limited range (two models) near the top of the market for which any substantial Australian demand existed. Within the power classes in which it has offered models, Chamberlain's prices in the sixties appear to have been somewhat lower than those of the three largest firms, although information is only fragmentary.[70]

Chamberlain not only suffered from a thin tractor line produced at only a few thousand units a year, but undoubtedly also from a lack of complementary equipment in a country in which the largest and best dealers have traditionally been those of Massey-Ferguson, International Harvester, and Ford (the latter's dealers complementing tractor volume with cars and trucks, as well as various implement lines). Chamberlain never produced its own engines and appears to have relied exclusively on Perkins both before and after the takeover by Massey-Ferguson. As was the case in North America, the small-volume firm could have been driven out of business by its giant competitors, but in addition to the cost of doing so, the red ink on Chamberlain's books might well have led to direct difficulties with the Austrialian government which was under constant pressure from Chamberlain to substitute a 37.5 percent tariff for the subsidy system. The importance the federal government attached to keeping Chamberlain alive was shown several times in the fifties and sixties when "bounties" were raised.[71] Avoiding any increase in the likelihood of adoption of the proposed tariff would probably have seemed a small price to pay for keeping Chamberlain afloat, given its rather modest market share (between 5 and 10 percent).

The hypothesis can be stated generally: When the government of a country is interested in the success of a domestic enterprise, and a definite proposal for substantial tariff protection is being put forward (presumably by the domestic firm itself), the importing firms may well be interested in maintaining the economic health of their domestic rival. This will be more true, inter alia, as the domestic firm's market share is small, economies of scale in the production of the industry's product are great, and imports are from areas of relatively low absolute cost.

The experience of the sixties, which apparently saw little, if any, expansion in its share, may have convinced Chamberlain that something drastic would have to be done in order to improve its fortunes. One obvious solution was to combine its operation with that of a firm able to augment its product line, and an association with Deere in 1970 probably came as no surprise to industry observers. A single operation was formed to handle all of the Australian activities of both companies, and Deere was to acquire 49 percent ownership of Chamberlain over a period of six years. Immediate plans called for both lines of tractors to be handled by a common dealership system.[72] The agreement had only a minor impact on concentration because Chamberlain's 10 percent share at best was combined with that of Deere which was certainly under 5 percent.

Concentration data for Australia are not sufficient to permit any conclusions about changes since the mid-fifties, other than that the role of the largest firm, Massey-Ferguson, probably declined somewhat, while the five-firm ratio probably increased in the sixties as Fiat rose to challenge and perhaps surpass Chamberlain. No exits from the market were recorded subsequent to the departure of the tiny post-war domestic firms and the Continental would-be participants discussed in the previous chapter.

THE ROLE OF COOPERATIVES IN THE
MANUFACTURE AND DISTRIBUTION OF
FARM MACHINERY

The U.S. Federal Trade Commission Report of 1938 confirmed long-standing farmer suspicions about the market power of the farm machinery industry by concluding that monopolistic tendencies in the production and distribution of implements were very strong, adducing as evidence International Harvester's ability to raise prices and maintain or increase profits during some depression years.[73] Partly in response to the general mood of agriculture in the late 1930s, 13 regional cooperatives (2 of which were in Canada) banded together to run the National Farm Machinery Co-operative which obtained the plant of an ill-starred town project of the depression and manufactured tractors between 1940 and 1952. The federation was in turn owned by local cooperatives which acted as retailers for the tractors and other manufactured and purchased machinery.[74]

After selling its plant to Cockshutt in 1952, the federation continued to market equipment into 1954, and one member coop did so until 1962, but the group was apparently unable to compete in a tight market. One respected student of the industry has suggested that salesmanship, credit, servicing, and trade-in evaluation were concerns for which the cooperatives' other activities poorly prepared them. Furthermore, cooperatives have a reputation for offering insufficient remuneration to attract high quality management. Nevertheless, some local cooperatives operated as retail dealers on standard terms throughout the period and were tolerated, if not enthusiastically received, by the rest of the retail trade.[75]

Cooperative production and distribution have been considerably more important in Canada than in the United States. The Canadian Cooperative Implements Limited (CCIL) both manufactures and sells implements in Western Canada.[76] It began distribution in 1945, and during the late sixties it sold about 5 percent of the total implements supplied in the Prairie Provinces: Alberta, Saskatchewan, and Manitoba.

CCIL was begun partly in response to a parliamentary report in 1937 which suggested cooperative action as a means of bringing down distribution costs, but the organization had problems finding a tractor source during the post-war boom. By joining forces with the National Farm Machinery Co-operative however, it persuaded Cockshutt to market through both organizations, while at least in Canada, only dealer trade terms were received, and the equipment was sold with the agreement that only list price would be quoted. White Motor Company apparently found the arrangement between Cockshutt and the Co-op unsatisfactory, and after the U.S. firm purchased Cockshutt, CCIL was forced to offer European machines; in 1970 it was importing Volvo tractors and combines and Deutz tractors to complement its own implement production which accounted for half its sales.

The distribution system of the cooperative altered considerably after 1945. At first, the plan was to sell for cash only, not to take trade-ins, and to sell through local cooperatives at a very small commission. Within a couple of years, the third element was dropped, a sales staff was acquired, and large sales agencies called depots were established; the goal of 60 depots, first set in 1951, was achieved by 1966. These outlets substituted for the hundreds of dealerships which many competing firms enfranchised in the same trade area. Simplified distribution helped produce enormous overhead cost savings. Between 1960 and 1966, selling, general, and administrative expenses were only 1.9 percent against 12 percent for the major companies. Trade-ins were allowed starting in 1952 and quickly became, in effect, the principal means of price competition.[77]

Although the earnings of the Co-op up to 1970 were very respectable by commercial standards, its market penetration was poor. The depots were apparently on a size which was adequate to realize all important economies of scale, service was reported to be very good, and there was little farmer complaint about distance to a depot. There was also a dividend which was not usually paid in cash but rather remained in the firm until the farmer's retirement—it was 12 percent in 1966. The present value to a farmer of such a dividend realized in 15 years (for example) and discounted at 10 percent is only about 25 percent of its nominal value, however, and many farmers regarded the dividend as a mere gimmick.

Apparently, generous trade-ins cum dividend were not enough to outweigh price and product line problems, and this is scarcely surprising. The Deutz tractors sold by CCIL in 1968 were up to 35 percent higher in list price than comparable models from other sources. Furthermore, there was widespread complaint about the inadequacy of power choices. (Deutz offered machines only up to 80 h.p. as late as 1968, and the largest Volvo was 90 h.p.) Finally, CCIL's switch from one tractor line to others undoubtedly did not increase confidence, although the loss of Cockshutt did not create any parts-availability problems because the machines were still sold by other dealers. Nevertheless, farmers might wonder if this would still be the case if, say, Volvo and the Co-op were to have a falling out. Furthermore, the successive marketing of three different combines in less than a decade could not have added to CCIL's image of stability.

In 1970 CCIL responded to its lack of marketing success with a bold new move; 30 percent was to be given off list price for cash purchases and an additional 1 percent a month up to a total of 42 percent was to be given for orders paid in advance (CCIL's purchase price is not known, but 40 percent off list price is probably about right). This plan seemed to deal with several problems at once. By shifting a part of what was hoped to be increased demand at least in part directly to the factory, CCIL could avoid the least attractive aspect of increased sales: the necessity for larger inventories on a limited capital base. To combat the trade-in loss problem it was proposed that the used machinery be disposed of by occasional auctions at which the farmer could either sell his own equipment or could have it sold for him for a small commission.

Doubts could still be entertained about CCIL competitive potential given its imported products. These came from low-volume producers in high-wage countries, and there is no indication that CCIL paid a particularly favorable price for them relative to wholesale outlets in other countries.

The Coopérative Fédérée de Québec enjoyed a 10 percent penetration of the total market for machinery in its province in the mid-sixties; it was the Oliver distributor in Québec, and it also began importing Renault tractors in 1960. Most of its trade was with local cooperatives, although one-fourth was with very small franchised dealers. It is probably fair to stress that this cooperative would warrant close inspection only for cautionary reasons because, despite the fact that farm machinery sales were a small part of its total business, it lost money on this line in nearly every year in the late sixties.[78]

By the standards outlined in Chapter One, it is doubtful that either Volvo or Renault should be characterized as ever "entering" the North American market. Half-hearted attempts by both firms to sell in North America in the fifties were completely unsuccessful, and their role appears to have been rather passive in their dealing with the cooperatives they have served. Neither firm by 1970 engaged in any other North American activity connected with tractor sales; further, their actual sales were miniscule. Volvo sales were under 400 in Canada in 1970 and Renault sales totaled less than 100 in 1969. (The Japanese firms selling in North America at the same time were also doing a small business there, but their participation was just beginning.) On the other hand, Deutz's entering into a supply relationship with CCIL seems to have been merely one step in developing a coverage of all North America. Outside of the Canadian Prairies and in the United States the company distributed to dealers through wholly-owned branches.[79]

Information on cooperative activity outside of North America is exiguous, but at least two generalizations are possible. First, in no market does their role in either manufacture or distribution seem to have grown substantially over time, and second (and presumably not unconnected with the first point), there is no indication that cooperatives on the whole have either provided better service or lower effective prices than their competitors. On the former issue, the French experience is negative. Although 10 percent of all French outlets by count in the mid-sixties were cooperatives, and they accounted for 30 percent of Renault's sales, only 15 percent were reported to keep spare parts or provide after-sales service.[80]

On the question of effective price to the final purchaser, the evidence is equally unfavorable to cooperative distribution. There is no evidence of the successful exercise of monopsony power at any level of distribution in any market. No material was found which indicated dispute between cooperatives and their suppliers or any attention to prices paid in other countries for similar equipment. Indirect evidence on the issue, however, is strong: dealer groups, especially on the Continent, attack cooperatives regularly as a matter of policy, but their attack revolves around special legal, especially taxation, advantages conferred by the state and not their superior bargaining power.[81] Furthermore, for a variety of reasons already outlined in the discussion of U.S. cooperative distribution, only CCIL showed evidence of effectively lowering the wholesale-retail margin by comparison with alternative channels. The cooperative which loomed largest in any market was the *Federconsorzi* of Italy in the early fifties, and of it a careful observer reported, ". . . it does not appear that this organization is succeeding in supplying the equipment used in agriculture on terms which differ appreciably from those of a free market."[82]

CONCLUSION: PATTERNS OF BEHAVIOR AND STRUCTURAL CHANGE

Full Lines and Exclusivity

As was predicted in Chapter Six, where North American full-line firms became important competitors in the various markets, they attempted to sell a broad product line, and their increasing success resulted in countermeasures by their principal rivals. In North America itself Ford was forced to expand its offerings, and the phenomenon is seen in the reactions of Renault and Deutz on the Continent and possibly also in Australia in Chamberlain's realization of its long-run lack of viability. The German experience even before the growth of full-line selling suggests confirmation of the hypothesis that exclusivity will be desired by the strongest firms whether or not full lines are important. The Italian case also confirms the hypothesis for the domestic market has been largely insulated from direct foreign influence first by finance discrimination and high tariffs and then by low domestic prices (see Chapter Ten), and, although American marketing appears not to have developed, exclusivity nevertheless became the order of the day.

Sales per Dealer

International differences in sales per dealer are obviously affected by a large number of factors, including geographical compression of tractor use, the number of brands sold, and the degree of concentration (which could, among other things, affect the degree of mutual dependence recognized in dealership proliferation or diminution) and the extent of exclusivity. Many of these differences, except for geographical compression, diminished over time, and it is therefore not surprising, in light of the discussion in Chapter Six, that in the late sixties there was a much lower rate of tractor sales per dealer in North America and Australia than in Europe. In the U.S. a typical dealer was selling about a dozen machines, in Canada somewhat fewer, and in Australia only about a half-dozen. In Britain and on the Continent, however, a typical dealership sold between 20 and 50 tractors, depending on the country and the firm.[83]

Distribution Expense as Product Competition

Dealer credit is a form of product competition which was already well established in North America before the war, and its use elsewhere is instructive. In Britain, where two North American-related firms dominated the market, it must have been determined early on that whatever increase in tractor sales might be induced by the provision of liberal dealer credit and the concomitant increase in farmer exposure to more machinery would raise costs more than it would raise revenues; thus it failed to gain any role in the industry. Elsewhere, provision of dealer credit was never non-existent (except apparently in Italy), but until Deere's lavish practices on the Continent in the sixties, the dealer always paid a penalty for the credit, which still might have been provided at favorable rates. Massey-Ferguson declared as late as 1970 that *only* in North America did it provide penalty-free dealer credit.[84] In sparsely settled Australia, it seems likely that a movement toward North American practice might have developed much earlier and faster than it did had the world industry's "working capital leader," Deere, played a more important role in the market. A manufacturer is probably most concerned about adequate exposure of his machinery in areas with a sparse tractor population. Therefore, Deere's new Australian role in the seventies leads one to predict that credit in Australia, where the display function must be as important as in North America, will ultimately be virtually the same as in the originating area, because such display (and perhaps retail credit provision) will doubtless appear to Deere to be the best way of increasing its market share.

The expense of North American distribution seems to be considerably in excess of that which would be necessary for the maximization of joint industry profits and reflects competitive activity in at least two respects: the rate of inventory turnover and the number of dealerships. If the industry suddenly abandoned floor-planning, it is very likely that most dealerships would keep perhaps no more than one tractor on display at all times rather than the three or four they now have, and it is extremely difficult to believe that total industry sales would be significantly affected. The extent to which inventory turnover would be increased by the abolition of all dealer credit cannot be known with certainty, but evidence that it would be significantly raised is provided in the contrast of two moderately good sales years in the U.S. and Britain (1968 in the

U.S.–1966 in the U.K.). Turnover of new goods inventories averaged 7.2 in Britain as against 2.0 in the United States for dealerships of approximately the same magnitude in sales.[85] These figures are not entirely comparable, as unit prices are lower in the U.K., and the composition of sales between the two countries differs somewhat. The most important difference is that combines, with the lowest turnover of all equipment in North America, are sold in the ratio of 1 to 9 or 10 against tractors in Britain but 1 to 6 in the U.S.[86] Nevertheless, the order of magnitude of the difference in turnover is not likely to be substantially accounted for by differences in product mix.

Furthermore, the number of dealerships could be cut drastically under a well coordinated cartel, and all of the remaining outlets would still be within easy driving distance of the farmer. It might be found profitable in some areas to "pull back" wholegoods display somewhat farther away from the farmer than would be optimal for after-sales repair parts and servicing, in which case, separate "service centers" would be necessary (possibly stocking parts for and servicing all makes). In the late sixties, John Deere was establishing a system which bore some superficial relation to this form of organization, but the motivation was very different. Deere had severely cut the number of wholegoods dealerships in previous years, not to increase economy in distribution, but to increase the quality of service for its machines (these issues were touched upon in Chapter Five). To further this end Deere began to set up satellite "service centers," owned and operated by private entrepreneurs, often John Deere wholegoods dealers, in adjoining areas. While the system could ultimately cut Deere's wholegoods handling costs, the declared aim of the policy was to win a massive amount of business away from competitors in initial sales by the implied promise of better after-sales service.[87]

The overall development of distribution and product lines seems to confirm the observation by Caves that international entrants often stimulate more vigorous competiton than would otherwise obtain,[88] and they have certainly hastened the demise of many highly inefficient producers. Nonetheless, against this must be weighed the outcome of the competition: a situation of heightened barriers to entry for surviving firms.

Predatory Pricing

We also discovered possible international dimensions to "predatory pricing." In North America and Germany such pricing may have been avoided in part to keep established firms from the hands of foreign rivals. Another circumstance in which a small firm may be kept alive with the acquiescence of its larger competitors, illustrated by the Australian case, is when there is the possibility of tariff retaliation.

Structural Change

There is a variously formulated hypothesis in the literature of industrial organization relating product differentiation and economies of scale to a rather high level of concentration. Bain,[89] has suggested that product differentiation typically leads to some firms being initially favored, and the implication of his argument is that economies of scale magnify the entrenchment of those firms. Bain's argument suggests that product differentiation alone, however, is often sufficient for a high level of concentration to develop. In a more pointed formulation by Caves,[90] an industry in which some firms develop high product differentiation advantages vis-à-vis newcomers and where economies of scale are important evolves at least a moderately high level of concentration and a modest number of sellers. This hypothesis is well confirmed by the experience of the tractor industry. Of the dozens of firms producing standard wheeled tractors in the early post-war years, only a handful remained by the late sixties, and every market taken individually did attain at least Bain's "high moderate" level by the end of the period. The pattern was not a simple one in which the rich got continuously richer, for, even in a closed economy, an innovation like the Ferguson system by a newcomer could be expected to shift demand in such a fashion as to reduce the level of concentration by some indices.

Furthermore, the logic of the argument does not allow for the place of international factors in influencing the behavior of domestic firms or their potential foreign competitors. The decline in top-level concentration in North America since the late fifties is certainly not one which could have been predicted on the basis of domestic factors alone. It should also be stressed that, had the distributional product differentiation barriers not been as important in North America as they were, the international impact on behavior and perhaps on concentration, might have been far more dramatic than it was. The case of France, where the "dealer dimension" is a less formidable barrier provides a sharp contrast. The dramatic drop in the share of the top four firms in the French market would surely not have occurred in the absence of competition from firms well established in other markets. Italy, on the other hand, seems to present a case of the confirmation of the hypothesis as one would expect it to apply in the context of a gradually improving product and great scale economies in a closed economy. Because of the original demand for simple "lugging" machines. Ferguson-system hydraulics did not have the immediate impact experienced in North America, Australia and the U.K. Instead, hydraulics were diffused slowly through the late fifties and early sixties with Fiat and Same getting a larger share of the market in virtually every successive year. Furthermore, the market was well protected from important foreign influence after the early fifties, first by discriminatory finance bolstered by high tariff barriers and subsequently by the ability of domestic firms to quote prices their EEC competitors apparently could not profitably match, while finance discrimination against all foreign firms and considerable tariff protection against U.K. firms was maintained.

World Concentration

As just suggested, one of the reasons for expecting considerable variation in the level and change of concentration in any individual market was the extent to which firms located in one market penetrated elsewhere; France and Germany represented this phenomenon most clearly. What of overall concentration? Here the hypothesis formulated as suggesting secularly increasing concentration appears to be borne out. Relative to total non-Communist sales, the share of the "big four"[91] reached its acme before other sources of supply were really established (over 70 percent in 1950). Since the mid-fifties, however, their role appears to have increased again. The shares given in Figure 1-1 show that while in 1954, the first of the "normal" demand years for North America, the share was estimated to be 59 percent, this had climbed to 63 percent in 1966, the last year for which confidently held estimates for all four firms are available. All indications are that it rose further in succeeding years. The five-firm measure grew even more rapidly because of the rise of Fiat; several firms were producing in the 20,000-unit range in 1954 (Fiat not among them), and the five-firm figure was probably about 62 percent; it was 69 percent by 1966.

Chapter Nine

The Product: Design Convergence and Competitive Behavior

It has been suggested in Chapters Two, Seven, and Eight that important differences among tractor lines offered by different makers did exist during the first 15 years or so after the war. The purpose of this chapter is to examine those differences, the reasons why they declined over time, and the extent to which by the early sixties most agricultural tractors sold over the Western world were very similar. Finally, an attempt will be made to examine the character of product competition (narrowly construed) in the industry.

DESIGN VARIATION AND CONVERGENCE

World War II and the immediate post-war years were a period of great uncertainty for the full-line producers serving the United States and Canada. They were suffering an intense attack from the Ford-Ferguson machine but remained unconvinced of the long-run appeal of integrally mounted implements, particularly on large tractors. Over the post-war decade, however, several of the major firms experimented with various forms of integral mounting. In 1954 Oliver introduced a very close imitation of the Ferguson system, and in the following few years, the system became universally adopted. The resistance of the established manufacturers to the Ferguson system has been widely denounced retrospectively by engineers in the employ of major firms themselves, and explanations usually turn on management and technical staff inertia and lack of imagination. In fact, it is additionally possible that part of the resistance to the new machines was born of the same motivation which made Detroit chary of building small cars. The smaller machines seemed to promise more work to be done with less equipment expense. The first response of the industry, reflected in its 1949 standardization of hydraulically controlled trailed implements, was to offer a sophisticated alternative to the Ferguson system. As the long-run superiority of integral mounting became obvious, however, various versions were

introduced by all firms, and the parameters of the implement and tractor interface were subject to formal industry standardization in 1959. By the late fifties
most major U.S. manufacturers were building essentially the same kind of
tractor, with similar engines, transmissions, and hydraulics. Deere held on to
two-cylinder engines until 1960, but it then introduced tractors with four and
six cylinders similar to those offered by the other large firms.[1]

In Britain, the complete domination of the Ferguson design was achieved
even more rapidly than in the United States, apparently because expanding
manufacturers had little vested interest in older designs. Ford's hold on the
market began a precipitate descent after Ferguson began producing there in
1946; David Brown also marketed a close imitation of the Ferguson machine as
did the Nuffield organization. Ford introduced Ferguson-type machines in the
U.K. in 1952, but International Harvester lagged behind, as it did in the U.S.,
and did not offer really sophisticated hydraulics until the late fifties. This presumably goes far to explain its poor performance in the British market.[2]

The rapid rise to predominance of Ferguson-type tractors in Australia was
similar to that in Britain, and despite it domestic fabrication, International
Harvester suffered an enormous competitive disadvantage until an imitation of
the Ferguson mounting technique was introduced in the late fifties.

On the Continent, as noted in previous chapters, a different situation prevailed. Rural poverty and small landholdings made machines less powerful than
those sold in the other markets far more important, and hydraulics of any kind
were regarded as something of a luxury until the mid- to late fifties. Most
farmers saw the new machines as "draft tractors to replace horses."[3] Nevertheless, in France International Harvester and Massey-Harris kept an edge in
product competition simply by gradually introducing features long standard elsewhere: simple hydraulic "lifts" which increased ease of implement use and
mobility but did not control working depth, multigeared transmissions, more
versatile power take-offs, and variable wheel spacing.[4]

In Germany, landholdings were far smaller than in France, and much of the
market was initially served by tiny one- or two-cylinder machines of a kind with
which the multinationals were wholly unfamiliar. Local designers experimented
with machines particularly suited to one-worker farms of only a few acres. A
lasting contribution was the Daimler-Benz "Unimog," something of a cross
between a tractor and a small four-wheeled drive pickup; this machine was introduced in 1952 and was still selling in 1970, although it seldom had more than
4 percent of the market.[5] Another Germany development, the tool frame,
positioned the driver behind an open frame which could be fitted with a truck
bed or bars for the mounting or various implements. This type of machine also
failed to capture more than a few percent of the market, and unlike the Unimog,
its small share diminished considerably during the sixties.[6] While the tool-
frames were often ingenious, it is probably not an oversimplification to say that
their ultimate weakness was that by trying to combine the functions which a

number of other machines would otherwise have to perform, they were relatively less effective at nearly every task, and, as average farm size increased, their modest attraction diminished.

The share of the small four-wheel drive units in Italy, about 10 percent of the market,[7] has already been discussed in the previous chapter. Even more common there has been a standard tractor with normal-sized, but power-driven, front wheels to increase traction. All of the domestic leaders produced most of their models with this simple four-wheel drive option.

Despite the role of specialized machines for some agricultural uses, most farm tractors sold everywhere by 1960 had become almost identical in basic design. This was one of the main reasons for the demise of the "two-line" policy pursued by Massey-Ferguson after the merger until 1957. Tractors of the same power were so similar when built to best practice design standards that to allow differences in underlying design between two lines was thought to mean that one would inevitably dominate the other (and economics of scale would be sacrificed), while two virtually identical lines under different names would smack of fraudulence (although this policy was considered for a time).[8] The same potential problem existed for Massey-Ferguson when it took over Landini in 1959, although in the late sixties two lines were still being sold. When a major retooling of the Landini line is considered, one would predict that the old two-line issue will once again have to be faced, and Landini's agricultural tractors will disappear in fact if not in name, as did White's Minneapolis-Moline line. For the same reasons, it is probably a safe prediction that Australia's Chamberlain will not survive as a distinct design beyond the life of the present production equipment, because prior to its alliance with Deere, it could claim almost nothing unique except the nationality of its stockholders.[9]

The international design convergence just outlined was in part both cause and effect of the advantages offered to the most important firms in the industry of international product standardization which offered the achievement of greater economies of scale through consolidated parts manufacture; this phenomenon will be explored in Chapter Eleven.

INDUSTRYWIDE PRODUCT STANDARDIZATION

In the U.S. the engineering committees of the Farm and Industrial Equipment Institute (the manufacturers' trade association) suggest standards for parts dimensions and other specifications to the American Society of Automotive Engineers, and they are often adopted.[10] Similar measures are taken by trade and engineering associations in Britain and on the Continent. The voluntary compliance of the industry is then required, and response in the U.S. and elsewhere seems to fall into two categories. Where the standards have improved the compatability between a tractor and complementary equipment, they have been widely adhered to, and the eventual world standards have often originated in the U.S.; e.g., dimensions and positioning of the two major sizes of the "three-point hitch" which grew out of the Ferguson revolution. Apparently firms all over the world, including the largest full-line sellers, fear the destabilizing impact of linking the fortunes of a whole range of machines in one line by making them incompatible with those of other makers. No similar incentive for non-"interfunctional" parts standardization is apparent, however, and adopted standards have been widely ignored even where parts are only trivially different.[11] The companies may be more than merely uninterested; extensive parts standardization could lead to a growth of competition from "to fit" outside makers in a product line, spare parts, where profit margins are relatively high.[12]

The reason that some standards are developed and promulgated but not subsequently adhered to by the firms sponsoring the standardization may well involve the differing views of standardization taken by a company's management ment on the one hand and its engineering team on the other. The engineers' view is apparently that standardization should reduce the "variety of components required to serve an industry, thus improving availability and economy for manufacturer and customer,"[13] and this implies that when there is no functional reason for parts of various manufacturers to differ, they should be standardized, and attention directed to other problems. Although engineers do not make the final decision about what to produce in any industry, their morale may be buoyed by allowing them to work for socially beneficial solutions. Once they have standardized a part, the problem is out of their hands.

PRODUCT COMPETITION

What sort of product competition would one expect to see in an industry of this kind, keeping in mind that it has been essentially a replacement market in North America throughout and become one rather rapidly in most other markets? It has been argued that price has been used only sparingly as a competitive weapon, most notably by foreigners dissatisfied with their role in a market. In a moderately concentrated national industry where differentiability is important, the logic of industrial organization analysis and the experience of many industries leads one to expect "for better or worse, . . . substantial product competition and rather little price competition."[14] This general observation, however, must be interpreted carefully in the light of the character of product differentiation discussed in Chapter Four. It was argued there that proven dependability and established distribution came close to exhausting the product differentiation barriers to entry into the industry. As basic design differences became almost trivial, increasing the relative dependability of the product could be a very real kind of product competition and perhaps even an effective one in the long-run, but it is not known to have been pursued with substantially differing success by any firm. The extent of product competition through distributional arrangements, including credit, has already been examined. What else remains?

It is perhaps instructive to contrast the product variation options open to the tractor-maker with those of the car manufacturer. Both face a replacement market, with short-run demand largely determined by disposable income, and a stable underlying technology in which most technical improvements are freely licensed among industry members, apparently for the purpose of increasing industry stability.[15] The differences between the two industries, however, are more striking than the similarities. The first point is that the essential design of the tractor makes it less amenable to styling manipulation, and even if styling were much more important in selling tractors than it is generally believed to be, appearance alteration for any purpose, including imitation, is both simple and inexpensive. When David Brown became associated with the former U.S. Ford distributors in 1964, the new sellers were uneasy about the dated-looking round hood and "hunting pink"[16] color of the line. Appearance modification was accomplished almost immediately, and the revised look, the first David Brown revision since the late fifties, still prevailed in the early seventies.

Especially given the ease of appearance alteration, there is considerable indirect evidence that styling within rather broad limits has not been of great importance in the industry. Ford's little tractors (a close imitation of the original Ferguson design) which were sold in North America in the fifties were augmented by larger models over the decade, but the physical appearance did not change significantly until a newly engineered line built to worldwide specifications was introduced in 1965. Massey-Ferguson, too, introduced a new line in 1965, and styling changes were marked, but the predecessor machines had looked virtually the same back to the early fifties. Neither firm subsequently made any substantial modification in appearance up to the early seventies, and in both cases the new "look" accompanied fundamental engineering changes and worldwide standardization. In spite of the rather minor importance of physical appearance, however, it is probably no accident that two of the North American firms' chief Continental competitors, Fiat and Renault, both introduced freshly styled machines in 1965, although in the Renault case, at least, important technical improvements accompanied them.[17]

While 1965 may have been a big model year for several of the major firms, this was not true for all, and really tight and virtually universal model cycles of the kind typified by the U.S. automobile industry have never characterized the tractor industry in any country. Deere's "New Generation of Power" of 1960 provided a handsome, well-engineered, and many would say long overdue— 37 years is a long time to produce essentially the same design—line in both North America and Europe [18] which came as no surprise, provoked no obvious and immediate response from competitors, and was still selling in essentially unchanged form in the early seventies. International Harvester's product conduct appears to have been more piecemeal than that of the other major firms, but over the years, it pared its three post-war North American tractor lines, McCormick-Deering-Farmall, McCormick-Deering, and International, down to two, Farmall and International. Furthermore, both remaining lines had very similar offerings because the original distinctions among the lines, based on differing regional farming patterns, became blurred by the increased versatility of tractor design and the opportunity thus afforded for realizing economies of scale. Comparable models in each remaining line usually had different axle and hydraulic packages and corresponded to machines sold by competitors as sub-models.[19]

With respect to international styling influence, the most that can be said is that the competition of the American firms probably prodded Continental firms into the production of machines reflecting somewhat more sophisticated industrial design than might otherwise have been the case. By the late sixties, however, all of the major Continental producers turned out machines every bit as modern-looking as their North American competitors.

If styling changes are both easy and not usually very important when unaccompanied by anything more fundamental, this represents an enormous contrast with the automobile industry, where lead times run to several years, are of great expense, and can greatly shift market share.[20] Another important contrast with the automobile industry is that while the sales of both are strongly dependent on income, the automobile market is complicated by the vagaries of "taste," while the post-war trend in tractor demand has been clearly toward secularly increasing larger average models in every market, and the only uncertainty is about the *pace* of that change. It is for this reason that the theory of "spatial competition" proposed by White for the automobile industry doesn't appear to hold for tractors. White argues that a certain kind of car (the major illustrations are "compacts" and "sub-compacts") won't be offered by one of the major firms until there is room for all of the "big three" to enjoy reasonable returns in that part of the market.[21]

In tractors, however, the issue could be predicted to be quite different even at higher levels of concentration and higher costs of low volume model proliferation than actually obtained because manufacturers could take turns offering the highest horsepower units, or some firm acting as "product leader" could typically lead the way with others following as the market broadened. One firm could "test the water" with a slightly larger model (often with an outside-purchased engine—this became less common with the development of engine "families" such as that introduced by Ford in 1965 in which the engine block and most ancillary parts can be adapted to a very broad spectrum of power output). The firm's success in subsequent sales could then help other firms to decide on the timing of their offering of higher power machines. The cost of introducing "top end" machines was usually minor enough not to be an enormous blow to the introducing firm if market acceptance was lukewarm, and the observed behavior avoided the extremely small unit sales for each firm which would have accompanied a simultaneous duplication of highest horsepower offerings.

Table 9-1 indicates that until the early sixties various firms led the industry in offering the most powerful machines, but then Deere, after it had taken over as the largest U.S. farm machinery producer and at about the same time as it took over the unit lead in tractors, became the firm which consistently offered the most powerful standard configuration farm tractors (this leadership continued into the early seventies). It must be stressed, however, that the picture is complicated by the fact that four-wheel drive tractors of various types are not shown in the table, nor is there any record of how many such machines were sold in any given year (the share was certainly not more than 5 percent by 1970). While Deere has maintained its lead in offering the most powerful machines for corn farming (where four-wheel drive is generally not suitable), other firms have offered four-wheel drive machines of considerably higher power, principally for use in wheat-growing areas. In 1970, both Case and International Harvester among the major makers offered machines of greater power than Deere's, as did several other firms.

MODEL PROLIFERATION

One obvious form of product competition would be the offering of many different models. The theory of industrial organization predicts that the higher the level of concentration and the smaller the number of sellers, all other factors held constant, the more cartel-like behavior will be. In this dimension of industry action, one would expect cohesion to result in only enough product variety to maximize industry profits. Given the difficult to estimate and probably modest, but still non-negligible, cost of offering models of different sizes, one would predict that a well-coordinated oligopoly would offer fewer models than a more competitive market and force the buyer to choose among them, assuming, as is reasonable given the actual offerings in each market, that this will virtually never result in a failure to buy any machine at all. How could this hypothesis be tested for the tractor industry? The most complete information on model offerings is available for the North American market, and there appears to be a significant increase in the number of basic models offered by the major manufacturers over time. Several of the firms offered only three or four basic models at the end of the war, while most of them were offering at least a half-dozen by the end of the sixties [22] (this latter period was one of much experimentation with four-wheel drive offerings, sometimes purchased from the outside, and these models are not counted). Is this proliferation a sign of increased product competition, moving the industry away from the set of models which would maximize joint industry profits? Schwartzman concludes that it is, but he neglects a number of important factors. Some of the product proliferation may well result from increased interfirm competition during the period, and concentration did decrease somewhat. Surely, however, a large part of the explanation for the increased number of models is the more varied structure of underlying demand.

Table 9-1. Largest Horsepower Size Offered for Sale, Wheeled Tractors, Major Manufacturers, 1949–69

	Allis-Chalmers	Case	Deere	Ford	International Harvester Company	Massey-Ferguson	Minneapolis-Moline	Oliver-Cockshutt
1949	27.6	22.3	51.0	26.4	22.2	61.4	48.8	45.0
1950	35.8	22.3	51.0	27.3	53.2	61.4	59.5	45.1
1951	35.8	22.3	51.0	27.3	53.2	68.2	59.5	45.1
1952	35.8	61.8	51.0	27.3	53.2	68.2	59.5	57.8
1953	45.4	64.8	51.0	40.6	53.2	68.2	59.5	57.8
1954	45.4	64.8	51.0	40.6	67.2	68.2	59.5	58.1
1955	45.4	64.8	67.6	46.9	67.2	68.2	68.5	83.5
1956	45.4	64.8	67.6	46.9	67.2	68.2	68.5	83.5
1957	45.4	64.8	75.6	46.9	67.2	68.2	68.5	83.5
1958	54.4	64.8	75.6	50.2	67.2	68.2	68.5	83.5
1959	54.4	71.0	75.6	50.2	81.4	63.3	78.5	89.3
1960	54.4	71.0	75.6	50.2	81.4	63.3	78.5	89.3
1961	54.4	80.6	84.0	50.2	81.4	101.0	78.5	89.3
1962	71.6	80.6	84.0	66.9	81.4	101.0	78.5	89.3
1963	71.6	80.6	121.1	66.9	81.4	101.0	101.0	89.3
1964	71.6	80.6	121.1	66.9	81.4	101.0	101.0	89.3
1965	103.1	80.6	121.1	66.9	94.9	101.0	101.0	105.8
1966	127.7	86.2	133.2	66.9	112.6	120.5	110.8	105.8
1967	127.7	101.8	133.2	66.9	112.6	120.5	110.8	105.8
1968	127.7	101.8	133.2	66.9	116.1	120.5	110.8	105.8
1969	127.7	101.8	133.2	105.7	116.1	120.5	111.0	132.8

Note: No four-wheel drive or propane gas models were included. Data are maximum belt or power take-off (PTO) horsepower at rated r.p.m. Highest horsepower tractor offered in each year is underlined.

Source: Barber, *Report*, p. 140.

The units offered at the high end of the market moved *pari passu* with increasing farm size and labor costs and made possible the increasing speed of major farm operations, while farmers often employed smaller, more maneuverable and cheaper (to buy and run) units for utility operations; these machines were similar in power to those demanded by a large and growing non-farm sector. (In 1970, the most popular size class for non-farm use was 50 to 59 h.p., while the most popular class for farm use was 90 to 99 h.p.) Furthermore, many of the "models within models" which proliferated even more noticeably than the number of basic models, and which Schwartzman records with alarm, may very well have contributed little to unit cost while at the same time increasing industry profits by making unnecessary many of the independent "conversion" units which previously had been sold outside the industry to adapt tractors for varying uses. Finally, looking only at the North American market, the efficacy of multiple model offerings as a competitive device appears to have been diluted by other factors. While in the late sixties, most firms were offering between six and eight standard configuration tractors, John Deere with six was pulling ahead of International Harvester offering nine.

Despite the rather inconclusive nature of the North American evidence, comparisons among markets are revealing. Compare the market having the highest level of concentration in the fifties with that having the lowest. Ford and Massey-Ferguson served British and many sterling area export markets essentially as duopolists during most of the decade, offering only one model each; by the end of the decade they offered two. In Germany the industry leader in 1957, Deutz, offered eight distinct models (not counting sub-models), all below 60 h.p., as were the British machines.[23]

Although multimarket strategies in design and manufacture came increasingly to complicate meaningful comparison over time, it appears that even in the late sixties, the somewhat lower level of German concentration, and particularly the greater number of sellers of conventional tractors by comparison with nearly all other markets, was contributing to greater product competition. In 1969 International Harvester and Deere were both offering six models in Germany between 25 and 70 h.p. by comparison with four in the home market, although the unit market in Germany was much smaller both for the firms themselves and overall. Massey-Ferguson, although exporting to Germany from France, was expanding its share rapidly and also offered six models by comparison with only four in North America and the U.K. The largest seller, Deutz, also offered six.[24]

Advertising

The thrust of the argument above has been to minimize the role of physical product competition in the North American market; the role of advertising, too, does not seem obviously above the joint-profit maximizing level, although there is some evidence of an attempt to increase market share on the part of the smaller firms. The amount spent by North American farm machinery makers, an average of less than 1.5 percent of sales revenue during the sixties (and there is no reason to believe that tractors are advertised significantly out of proportion to their contribution to revenue) is quite modest by almost any standard.[25] An examination of foreign media suggests that a rather low level of advertising prevails there as well.

Despite the fact that the range of advertising expenditure for the major firms as reported by the RCFM was only 1.2 to 2.5 percent in 1962 and .8 to 2.4 percent in 1966 (the firms are unidentified), materials on advertising by firms in agricultural publications for 1970 strongly suggest a pattern by firm size within the range (which itself may have expanded by 1970).[26] The 1970 figures are only for a few months, so they do not allow annual expenditure-revenue comparisons, but they strongly suggest that Case, Allis-Chalmers, and White were doing much greater advertising relative to sales than were Deere, International Harvester, and Massey-Ferguson. An examination of the content of the advertising reveals that the smaller full-line firms, as one would expect, were heavily promoting their powerful tractor models. The payout from this effort is far from certain, however. For the period of 1962 through 1966, Martinusen and Barry state that "the difference in rates of spending on advertising did not correlate with changes in market share by company."[27] Indeed, only one episode has been recorded in which advertising was believed by those involved to have had an identifiable impact on sales.[28]

It has been stressed in previous chapters that the product differentiation disadvantage suffered by unknown newcomers is principally fought by selling at prices considerably below those of established firms. Nevertheless, entrants almost necessarily spend vastly more on advertising relative to revenue than established firms, although that expenditure, as one might expect, is not broadly based across the agricultural media but is focused on the trade press and therefore on potential dealers.

MARKET SHARE VOLATILITY

In the light of all the foregoing material, it is perhaps surprising to find market shares for tractors varying as much as they do in North America, where aggressive entrants have had little direct impact.

Overall volatility is suggested by RCFM figures which, for the eleven-year period of 1957 to 1967, give the difference in market share between the best years and worst years for the major firms except Ford and Allis-Chalmers. They were Massey-Ferguson, 9.4 percent; Deere, 5.4 percent; International Harvester, 7.9 percent; Case, 8.2 percent; and White (and predecessor companies), 5.2 percent.[29] The Massey-Ferguson difference is 41 percent of its 1965 Canadian market share.[30] Canadian market shares are not available for the other firms, but as a percentage of their estimated North American market share in 1966, the figures are Deere, 26 percent; International Harvester, 36 percent; Case, 132 percent; and White, 67 percent. Deere's relatively low volatility is noteworthy, as is the greater volatility of the shares of the smaller firms which may be due in part to the vagaries of the market acceptance of new high-horsepower machines upon which these firms have concentrated their efforts.

What are the explanations for this volatility? It has been stressed that price has been used as a competitive weapon only to a limited extent and only by the smaller firms, apparently with the indulgence of the larger. Furthermore, such differences have only been documented for the late sixties; inspection of prices for earlier years indicates that, if anything, they were closer together for comparable models. Another possible explanation is that tractors may vary in popularity with the strength of other machines in a company's line. We have already seen, however, an extreme example of lack of transference in Massey-Harris's post-war claim of half the North American combine market at the same time as it was selling only 5 percent of all tractors. More recent material collected by the RCFM also indicates little connection between a firm's performance in tractors and other major implements and certainly not with the latter influencing the former.[31]

One factor influencing North American market share over time has been the failure of some companies, particularly Ford until the late sixties, to keep up with the horsepower race (see Table 9-1). Another factor which is probably more important but is very difficult to assess, in part because of the lack of accurate year-to-year market share information, is the quality control of the major firms. Industry lore has Deere in North America as an almost faultless performer in this regard. It first produced and refined one product line for decades and then introduced another, the "New Generation of Power," in 1960 with an almost complete absence of engineering or manufacturing defects. Other Deere tractors virtually from the same drawing boards but produced in Germany, however, were plagued by the most severe quality problems in the company's history and help explain the firm's uphill fight there after 1960.[32]

Deere's experience in North America can be contrasted with that of Massey-Ferguson which had a reputation in the industry in the late fifties for noticeably poorer quality control than some of its rivals.[33] There is no question, however, that by the sixties these problems had been completely overcome and that Massey-Ferguson's products were virtually as trouble-free as those of any of its competitors. What one does not know, however, is the "decay" factor in buyers' perceptions relating to the earlier difficulties. A closely related problem, difficulty with a particular model, is illustrated by the experience of Ford. In 1960, the year of Deere's impressive line change, Ford introduced its "6000," "perhaps the most complicated farm tractor ever built."[34] Unfortunately, the machine suffered from such severe lack of pre-sales testing that practically every unit had to be recalled, an action which could scarcely have left unaffected the sales of other models in the line—perhaps for years after the "bugs" were taken out of the 6000.

Except for Deere's apparent supremacy, however, the overall "quality" of brands sold in North America does not appear to follow any generally recognized pecking order.[35] We have virtually no information on the relative reputations for quality of brands sold in other markets.

The only large firm in the world industry which seems to have lagged systematically behind the others in both technology and machine size offerings appropriate for the home market has been Renault. It was slow to develop the hydraulic controls and attachments which became increasingly critical from the mid-fifties onward. The continual refinement of such features is not particularly challenging, and has been met by firms of greatly varying size, including many producing far fewer tractors than Renault, but keeping pace does demand ongoing attention, and Renault seems at various times not to have adopted innovations quickly enough to avoid losing market share.[36] This has also been true with respect to continually revising highest power offerings upwards over time; its market share losses in the late sixties seem to have stemmed mostly from this failing. In 1967, Renault, with a product line freshly engineered only two years previously, was serving only 60 percent of the French market for lack of high-horsepower machines.

When one is attempting to explain market share volatility, all changes in "product," not just those related to the physical properties of the machines or advertising, must be considered. Quantitative or qualitative changes in the dealer network of a firm, perhaps a more important competitive weapon than the tractor line itself, would tend to contribute only gradually to changes in market share and not to short-term volatility, but other "product" parameters could cause more rapid shifts in share. One example was the temporary differential of dealer and retail credit extension in the late fifties discussed in Chapter Eight. Another product parameter with short-run potential would be differing warranty provisions. In fact, however, warranties within North America (and *within* most other markets) have seldom varied significantly. Deere, however, did begin a round of warranty adjustments in the industry in 1969 by extending the warranty on some parts beyond what had been a virtually universal one-year standard over nearly the entire post-war period.[37] In this area, as well as most other areas of non-price competition, Deere has led the North American industry. In credit extension, its leadership has been worldwide.

Chapter Ten

International Pricing Patterns and Buyer Responses

Chapters Seven and Eight discussed the use of price as a competitive device as far as information allows; this chapter will examine three other important pricing issues: the development of domestic prices in the various markets over time, the structure of prices among machines of various power classes by country, and international price differences. The chapter will conclude with an examination of the reactions of dealers and buyers to the wide international dispersion of prices for similar equipment.

INTERNATIONAL PRICE COMPARISONS OVER TIME

It is difficult to compare prices within a single market over time because features are improved and in some cases move from options to part of the basic machine. Even more difficult is the international comparison of prices because of the insufficient specification of what is included in the base price for tractors sold in different countries, even when they are nominally the same model. Although open price publication is not practiced in the United States, quite detailed information is available on the suggested retail prices of all makes and models through the dealers' association's *Official Guide*. This guide applies to the United States only, although most of the major firms determine their Canadian prices by applying the Canadian-U.S. exchange rate, either net or gross of dealer discount, to suggested retail prices. Firms with special price lists for Canada (notably Massey-Ferguson) appear to sell somewhat more cheaply there, but the differences are seldom more than a few percent.[1] In Britain, Italy, France, and Australia some price information is freely available to farmers in agricultural publications, but unfortunately for purposes of international comparison, accessories included in base prices are often not recorded.[2] German price data have never been commercially published and are centrally collected only by the

Bundeskartellamt and even then only when firms establish "recommended prices," something German manufacturers had become reluctant to do by 1970.

As is the case in the automobile industry, the most important price for tractors is the dealer price, because very few sales in any of the study markets have been made at full list prices since the post-war boom. The discounts allowed to dealers in the late sixties varied from a high of 27 percent in North America to only 16 percent in France; the major firms in all countries provide similar discounts for dealers of similar volume within each market.[3] Nevertheless, companies sometimes change dealer prices even after the equipment is at the dealership in order to clear inventory for new models.[4] As in the automobile business, price discrimination among customers within the same market takes place in part through the valuation of machinery taken in on trade.

In a situation in which markets are effectively isolated, the price for the same (or a very similar) product might be lower in one market than in another for a number of reasons: 1) production costs for the price-setting firms might be lower either because of scale economies (due to the size of the market and/or the market share of the leading firms) or absolute cost advantages; 2) non-production costs might be lower; 3) elasticity of demand for the industry's product might be higher; 4) barriers to entry of domestic firms might be lower; or 5) lower concentration might diminish the extent to which any degree of inelasticity might be effectively exploited (and perhaps economies of scale realized). Actual price differences would result from the net impact of all of these influences. In the incompletely isolated situations actually observed in the tractor industry, the question is further complicated by the fact that a national industry may be tempering its prices on all or some models of tractors in recognition of potential entry from abroad. While it is impossible to fully explain the level of prices in any one of the markets relative to the others without far more information than is available, a careful international examination of comparable machine prices over time is important for purposes of understanding a vital area of the world industry's behavior and performance.

Figure 10-1 is the result of a careful attempt to compare international dealer price of tractors in two important horsepower classes on a dollars-per-horsepower basis. The smaller of the two machines is in the 30 to 35 h.p. range and the larger in the 45 to 55 h.p. range, and their prices are traced over the period from 1955 to 1970 at prevailing exchange rates. At the beginning of the period, an overwhelming number of tractors sold in North America, Britain, and Australia were in these power sizes, while a majority of Italian and French, and the vast majority of German, machines were of lower power than the smaller of the two machines illustrated. By the end of the period, machines within the range of the sizes illustrated accounted for fewer than half of machines sold in North America, Britain, and Australia but for the vast majority of those sold on the Continent. (For a detailing of the comparisons made, see Appendix).

Bearing in mind that neither the exact quality of the machines nor the accessories are completely comparable, the time series comparisons are nevertheless

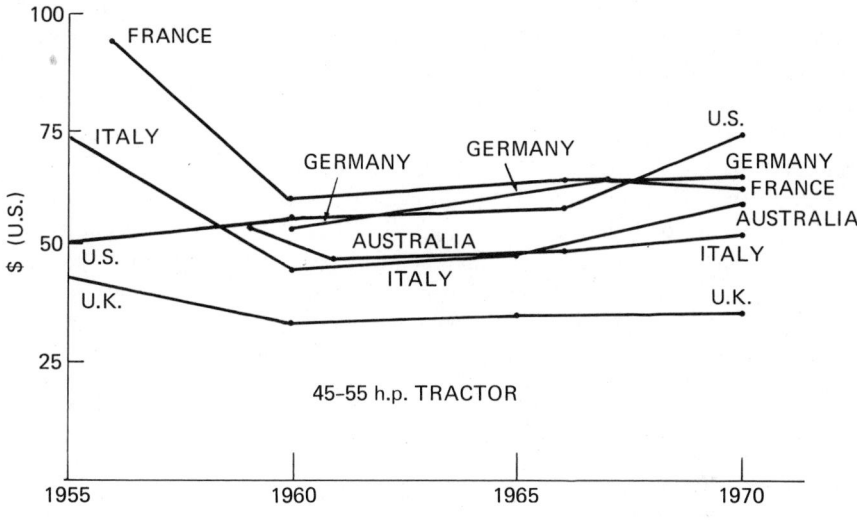

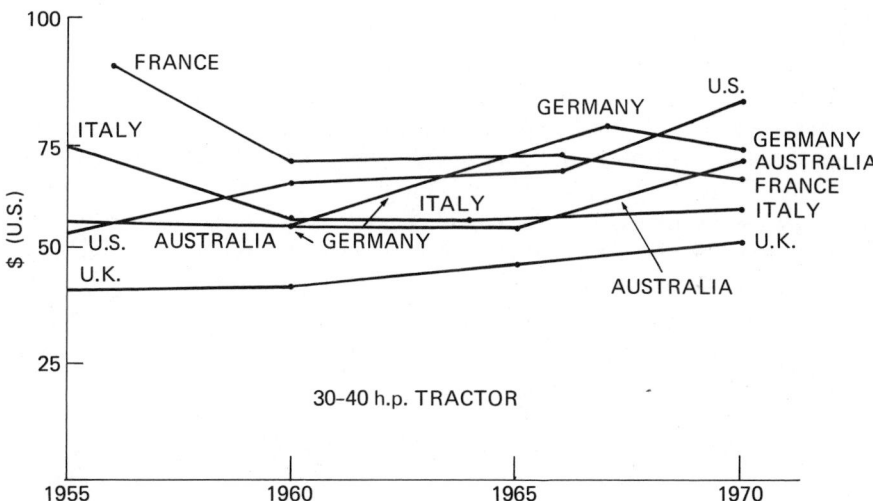

Figure 10-1. An International Comparison of Price per Horsepower at Prevailing Exchange Rates of Two Sizes of Tractors—Selected Countries and Years

Source: See Appendix 1.

very revealing. The United Kingdom has enjoyed vastly lower prices over the period shown; a tendency for home prices in dollars to creep upward was at least temporarily stayed by the devaluation of 1967. Figure 10-1 also presents the first evidence offered in this study that Italy has been a low-cost source of supply. Part of the wholesale price difference may result from low non-production costs in Italy, certainly less than 20 percent of the total, which in turn might reflect the lower level of wages and salaries which one suspects is a principal reason for the apparently low level of production costs. The Italian makers may also have realized unusually low profit rates, although there is no evidence that this was the case. Italian prices are lower than those of the other countries except the U.K., sufficiently lower that a strong *prima facie* case exists for believing that the difference reflects primarily production cost advantages. The relative stability of Italian prices during the sixties, despite substantial inflation and an unchanged exchange rate, may be attributable to the increasing realization of scale economies by surviving firms. Nevertheless, the price level as early as 1960, when even Fiat was producing only about 15,000 units per annum and other manufacturers were producing at much lower volume, suggests that Italy has been a country of low absolute cost.

The French industry's international prices benefited enormously from the large devaluations of the late fifties, totalling over 30 percent, and again from the 11 percent devaluation of 1969. From being a country with very high prices in the fifties, France became one with moderate prices in the sixties.

Although no usable 1955 prices are available for Germany, it is known that they were between those of France and the U.K.[5] German prices in the sixties are difficult to interpret. There was upward pressure from internally rising input prices (although the rate of inflation was scarcely different from that of France or Italy), exacerbated by the 4 percent revaluation of 1961 and the 8 percent revaluation in 1969. Furthermore, Deutz, the usual German price leader, apparently chose to pursue a price policy in the sixties which minimized pressure on its smaller domestic rivals, and the erosion of market share to imports was rather modest. Nevertheless, Massey-Ferguson's continually increasing market share and Fiat's entry into the market could scarcely have been reassuring to Deutz, and the impact on prices of free intra-EEC trade was clearly beginning to tell by the end of the decade. International prices in the lower h.p. class dropped between 1967 and 1970 and did so much more sharply when expressed in domestic currency.

The relative position of the U.S. industry by the standard used here is one of marked deterioration over the period. From being the country which had the lowest prices in 1955, except for the U.K., the U.S. came to have the highest prices in 1970.

It should be noted that tariff barriers contribute little to an understanding of the international variation in prices. The relation between tariff height and price level as measured by rank correlation is very slightly positive for some periods and tractor sizes and negative for others.[6]

One would surmise on the basis of the price material just presented that companies selling essentially the same tractors in different markets have done so at very different wholesale prices, and this has certainly been the case. Figure 10-2 is an attempt to put some price and cost data into cross-sectional perspective for

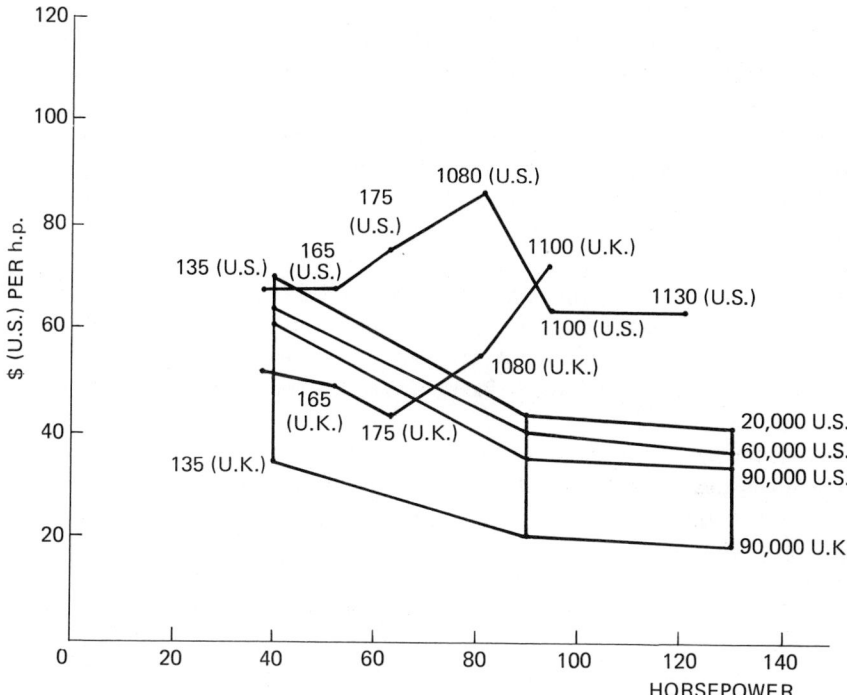

Figure 10-2.   Estimated Tractor Costs at Three Output Levels, U.S. and U.K., Compared with the Wholesale Prices of Models Sold in the Two Countries by Massey-Ferguson[a]

[a]Prices 1970; Costs 1968.

Source: Costs: Estimates developed in Chapter Three. Prices: N.F.P.E.D.A. *Official Guide,* Spring, 1970; *AMJ,* various 1970 issues.

a broad range of power sizes for two markets. The vertical bars with the straight lines connecting them are the estimated per-horsepower costs of tractors produced in 1968 at the relative output mix discussed in Chapter Three, where both North American and British costs for different size tractors were estimated. Wholesale prices, indicated by the other two lines in the diagram, are those of Massey-Ferguson and are for models sold in both the United Kingdom and in North America. The prices recorded are for a period two years later than that for costs, and the latter should undoubtedly be several percent higher; but the costs are rough estimates only, and the conclusions drawn here are not sensitive to either their exact level or their precise change with increasing machine size.

The first point to note is that at North American cost levels, the smallest machines could not be produced profitably by new plants. This would be true even if there were no non-plant costs, but the latter totalled over 20 percent of all costs for small firms. Secondly, for the first three models in order of increasing size, the price per horsepower declined in Britain, but increased in the United States, although the machines are only "similar," because some locally sourced parts were used on the tractors sold in North America. The fourth model, the 1080, shows a rise in cost per h.p. in both countries, but it was only slightly higher in Britain, where an 18 percent duty had to be paid on most of its value (the engine excepted) because it was produced in France, while all parts entered the U.S. duty free. The next model compared is the 1100 which sold for a higher price in Britain, closely approximating the additional transport and duty charges from Massey-Ferguson's Detroit factory, its only point of production. The 1130, also of U.S. manufacture, was not even sold in Britain in 1970 and is included to illustrate the gap which was sustained between price and cost in North America as horsepower increased.

That the pattern of declining cost per horsepower reflected in price decreases was not peculiar to Britain is suggested by Figure 10-3 which shows the wholesale price per horsepower of a number of different makes of various sizes sold during 1967 and 1968 in the different markets. In virtually all cases in Britain, Australia, and France, the "outliers" represent machines imported from the United States; although all of this U.S. equipment sold at higher prices outside of North America than at home, the additional margin was often not large enough to cover the full additional cost of transportation and tariffs.[7] These machines accounted for a negligible share of the market everywhere except Australia where they were still only a few percent of all units sold.

How can the level and structure of North American prices be explained? Some considerations can be advanced, but their relative importance is difficult to assess. First, production costs during much of the period have undoubtedly been higher than in several other areas. Second, it is undoubtedly true that the distribution system prevailing in North America is much more expensive than that prevailing in all other areas except Australia. Third, the level of concentration in North America is high enough to allow mutual dependence in price determination to be fully taken into account by all significant participants. Finally, the barriers to new competition are not exceeded in any other market except perhaps Australia.

The structure of price-cost margins suggest other considerations. One can conceive of different margins on different sizes of equipment resulting solely from a differing perceived threat of foreign entry into different parts of the market, while all parts faced similar elasticities of derived demand. Alternatively, higher margins on larger tractors could be fully explained by systematically less elastic derived demand in large-scale agriculture of various kinds owing to purely technical relations among purchased inputs. For example, the ability of high-speed tractors to do a great deal of work in minimum time may provide an important insurance factor in large-scale farming where the use of sophisticated

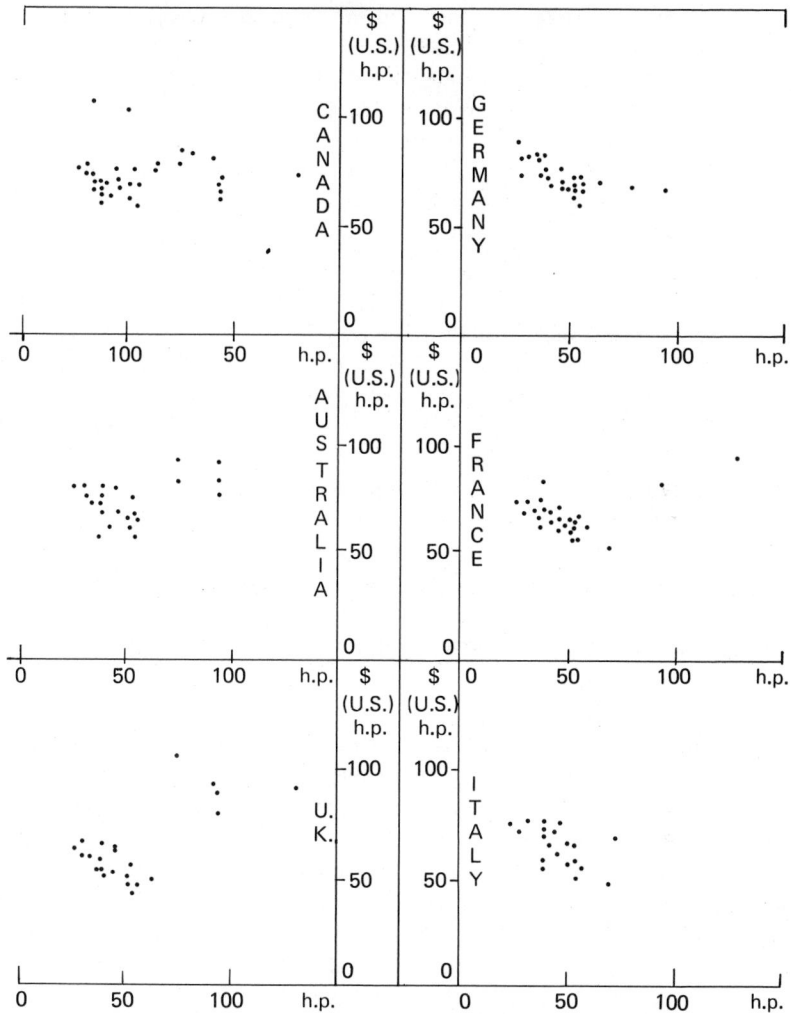

Figure 10-3. Wholesale Price per Horsepower—Selected Models and Markets, 1967 or 1968

Source: Calculated from Barber, *Prices,* pp. 116–142.

herbicides, pesticides, and fertilizers sometimes demands very precise timing. A third consideration turns on the nature of the farming enterprise in North America. It must be kept in mind that a typical farmer, even on a large farm, does not simply hire all inputs at fixed prices but provides a very important part of his own "labor force," It is reasonable to regard speed of operation and length or intensity of a work day as critical components of the farmer's living standard.

It is very likely that increased speed and shorter work days are highly income elastic and become more price inelastic with income. Net income, farm size, and the size of tractor purchased are highly intercorrelated and the higher margins on larger tractors may function as a form of price discrimination based implicitly upon the economic status of the farmer.

The cotton, wheat, and feed grains sectors of American agriculture buy the lion's share of large tractors (over 60 horsepower in the late sixties), and in these sectors the activities of the federal government have made a powerful contribution to increasing incomes. In evaluating the programs for these crops in the late sixties and early seventies, D. Gale Johnson estimated that, although market prices were increased very little by the government programs, direct payments connected with the production of these crops, accounting for between 80 and 85 percent of all government agricultural payments, raised net income in the three sectors by between 1 and 2 billion dollars annually. Furthermore, the payments were heavily skewed towards large farms which buy mainly large new machinery; on a large farm, government payments typically added several thousand dollars to family incomes already far above the national average.[8]

It is entirely possible that all three factors just noted contribute to the structure of price-cost margins in the North American industry.

INTERNATIONAL PRICE DIFFERENCES FOR SIMILAR EQUIPMENT

The substantial wholesale price differences between markets illustrated in Figure 10-2 are part of a general pattern. It has been common throughout the period to find virtually identical models being sold at much lower prices in the U.K. than in most other countries. Consider first, machines wholly produced in Britain and sold elsewhere. The most arresting comparisons are those between the U.K. on the one hand and North America and Australia on the other, because there has been no tariff to protect the domestic markets of the former colonies. As Tables 10-1 and 10-2 demonstrate, lack of tariff protection has not prevented huge differences among wholesale prices—ones which could be assigned only to a very minor extent of transportation costs. (See pp. 72-73). In the North American case, the machines compared were imported to complement domestically produced machines of other sizes or engine types. This motivation also applies to International Harvester in Australia.

Similarly dramatic are comparisons based on "functionally similar" machines which combined some components from the same plants with others sourced from the country of sale but which were nonetheless manufactured or purchased to world-wide specifications. Table 10-3 gives the relative dealer prices for the two firms engaging in such production for sale in the U.K. and North America; Massey-Ferguson engaged in the practice after the late fifties, and Ford after the mid-sixties. Expert opinion is that the quality of machines built in the U.K.

Table 10-1. Examples of Wholesale Price Differences: The U.S. and the U.K. (U.S. Dollars)

	Make and Model	*U.S. Price*	*British Price*	*Difference*	*U.S./U.K. Ratio*
1956	Fordson Major	$2,025	$1,391	$ 634	1.46
1962	IH B 275	$2,134	$1,276	$ 858	1.67
1964	IH B 414	$2,262	$1,544	$ 718	1.47
1964	County 4 (Ford)[a]	$5,058	$3,610	$1,448	1.40
1970	County 6 (Ford)[a]	$8,572	$5,878	$2,694	1.46

[a]County tractors are sold outside of the U.K. by Ford.

Source: National Farm and Power Equipment Dealers Association, *Official Tractor and Farm Equipment Guide*, St. Louis, Mo., various semiannual issues. *Farm Implement and Machinery Review, Power Farming, Agricultural Machinery Journal*, various issues.

Table 10-2. Examples of Wholesale Price Differences: Australia and the U.K. (U.S. Dollars)

	Make and Model	*Australian Price*	*British Price*	*Difference*	*Australian/ U.K. Ratio*
1956	Fordson Major (basic)	$2,156	$1,180	$ 976	1.82
1958	International B 250	$2,035	$1,347	$ 688	1.51
1961	David Brown 850 "Implematic live-drive"	$2,271	$1,450	$ 821	1.57
1961	David Brown 950 "Implematic live-drive"	$2,465	$1,639	$ 826	1.50
1963	David Brown 990 "Implematic live-drive"	$2,878	$1,697	$1,179	1.70
1970	David Brown 990 "Implematic live-drive"	$3,345	$2,385	$ 960	1.40

Source: *Farm Implement and Machinery Review, Power Farming, Agricultural Machinery Journal*, various issues; *Producer's Review*, various issues; *Power Farming in Australia and New Zealand*, 1970-71.

and North America does not vary in any significant way.[9] Comparisons were made on the basis of the writer's examination of English and U.S. price lists with careful attention to major options (transmission, steering, power-adjusted wheels, etc.). For 1970 two figures are given for both British and North American prices. The figures which show the wholesale prices as closer together are those of the RCFM. The discrepancy between this writer's British figures and those of the RCFM must result from an RCFM attempt to add minor features to the British machines to make them comparable with the North American tractors

Table 10–3. Examples of Wholesale Price Differences: North America and the U.K. Massey-Ferguson and Ford—Selected Models and Years (U.S. Dollars)

		Massey-Ferguson			
Year	Model	U.S. Price	U.K. Price	Difference	U.S./U.K. Ratio
1960	35	$2,216	$1,346	$ 870	1.65
	65	$2,557	$1,710	$ 847	1.50
1962	35	$2,213	$1,430	$ 783	1.55
	65	$2,927	$1,902	$1,025	1.54
1964	35	$2,420	$1,619	$ 801	1.49
	65	$3,215	$1,971	$1,244	1.63
1966	135	$2,540	$1,860	$ 680	1.37
	165	$3,519	$2,170	$1,349	1.62
1967	135	$2,682	$1,860	$ 822	1.44
	165	$3,538	$2,170	$1,368	1.63
1970	135[a]	$3,007	$1,848	$1,159	1.63
	135[a]	$2,573	$1,993	$ 580	1.29
	165[a]	$4,069	$2,116	$1,953	1.92
	165[a]	$3,507	$2,546	$ 961	1.38
		Ford			
1966	4000	$2,799	$1,996	$ 803	1.40
1967	5000	$3,970	$2,449	$1,521	1.62
1970	2000[a]	$2,608	$1,618	$ 990	1.61
	2000[a]	$2,447	$1,820	$ 627	1.34
	3000[a]	$2,850	$1,820	$1,030	1.57
	3000[a]	$2,681	$2,002	$ 679	1.34
	4000[a]	$3,808	$2,183	$1,625	1.74
	4000[a]	$3,637	$2,564	$1,073	1.42
	5000[a]	$4,619	$2,409	$2,210	1.92
	5000[a]	$4,419	$2,737	$1,682	1.61

[a]For an explanation of the difference between two prices for the same model, see text.
Source: National Farm and Power Equipment Dealers Association, *Official Tractor and Farm Equipment Guide*, various semi-annual issues. *Farm Implement and Machinery Review, Power Farming, Agricultural Machinery Journal*, various issues.

selected for comparison. The lower RCFM prices for the North American models may result in part from the removal of some features and also from the fact that the prices are from Canadian price lists which have simply been converted into U.S. dollars at the prevailing exchange rates; the latter factor alone may account for several percent of the difference, because Massey-Ferguson in particular does often sell slightly more cheaply in Canada than in the U.S.[10]

While the RCFM machine comparisons are certainly more accurate, the question remains: How is the North American purchaser offered his more feature-laden "basic" machine. Would the farmer not often purchase a cheaper but "functionally similar" machine if presented with a clear choice? Given the floor-planning practices of the industry, the dealer has some incentive to move the equipment he actually has in stock, and this may help to explain why basic machine and option suggested prices are sometimes not known by prospective purchasers; there is even evidence that some farmers are initially quoted prices above suggested retail price.[11]

With this in mind, it is unlikely that Bain's observation concerning market conditions in the early fifties that "the farmer is a better informed and more price-conscious buyer than the automobile purchaser"[12] would have been a tenable position in later years, especially after "sticker pricing" became mandatory for automobiles in 1958. What appears to be the case is that there is great variation among farmers in the degree to which attention is paid to features and price comparisons among all sizes of farming operations, despite the fact that price-cutting by U.S. dealers in the late sixties resulted in an actual realized margin only 11.1 percent of new machinery sales by contrast with the 23 percent nominal margin and average volume bonus of 4 percent.[13] On many large farms (2000 acres or more) realized dealer margins were 6 percent or less, according to a survey conducted in the Midwest in 1969. The survey pointed out, however, that increasing bargaining power of farmers rising with acreage was far from a uniform pattern. Many large farms were run by persons "who have gradually expanded their operations over a considerable period of time without adjusting their purchasing practices or trying to improve their buying skills.[14] The researcher who conducted the survey also points out that managerial skills will probably improve and become more uniform among the largest farms with time. In any case, the survey evidence does not contradict the point made earlier about the high income and low price elasticity of large fast tractors. Such an argument scarcely means that the larger farmer will not recognize that, as a large purchaser, he may have considerable bargaining power vis-à-vis any particular dealer. The leverage exerted is confined almost always to the retail margin, however, while our earlier discussion concerned wholesale margins.[15] Nonetheless, such bargaining power could create dealership problems for a manufacturer in an area in which its dealers face predominantly powerful buyers.

THE NORTH AMERICAN PRICES OF MACHINES
SOURCED FROM INDEPENDENT FOREIGN FIRMS

In light of the above material it is scarcely surprising that the North American wholesale prices of the machines imported by the small U.S. full-line firms to sell under their own names have been vastly higher than wholesale prices in the country of origin. In 1961 the David Brown "850," sold as the Oliver "500," was $700, or 49 percent higher in the U.S. than in Britain, the "Oliverized" Fiat "415" of 1966 had a wholesale price premium of $670, or 40 percent higher in the U.S. than in Italy, and Allis-Chalmers' bottom-of-the-line "160," sold in France as the Renault "56," cost the U.S. dealer 27 percent or $651 more than his French counterpart in 1970.[16]

BRITISH PRICE CHANGES AFTER 1970

Price information, however limited for some countries, has been collected for this study for all markets only through the first price lists issued for 1970. The RCFM collected no time-series data, and had it done its elaborate cross-sections for 1971, it would have found less dramatic but still substantial price differences between the U.K. and North America for similar machines.

While North American prices advanced only slightly, Table 10-4 shows the increase in British home prices for the 14 months after May of 1970. This price explosion was the result of an unknown mixture of suppressed price rises during the regime (1966 through 1970) of the Prices and Incomes Board,[17] and general inflation in Britain which saw the consumer price index rise nearly 10 percent between the end of June 1970 and the end of June 1971. Inflation at the rate of the first half of 1971, 11.5 percent, was not sustainable at the prevailing exchange rate. Unfortunately, the international price situation became difficult to interpret after the devaluation of the dollar because of the subsequent inability of the major trading nations to re-establish something like a permanent system of exchange rate adjustment and the unusually high rate of inflation in most countries. Chapters Twelve and Thirteen will nevertheless discuss the probable future course of relative international costs and prices in the industry.

THE RESPONSE TO DIFFERING
INTERNATIONAL PRICES

The persistence of lower prices in one market than prices in another invites arbitrage, unless transport costs or tariffs make this unprofitable, and, although one of the central conclusions of this study will be that price discrimination against North America coupled with arguably excessive selling costs by the major companies is possible largely because of massive U.S. farmer ignorance of, or confusion about, the facts, not all dealers have remained in the dark. Despite both British dealer contracts forbidding export or sale of machines for export and the practice of requiring British customers to sign an agreement not to sell for export within one year of date of purchase, some English dealers and farmers got around these obligations and systematically sold to dealers in the United States.[18]

The U.S. dealers, mostly along the Eastern seaboard, who did the "bootlegging" caused quite a stir in their market areas but apparently found it difficult to get machines in substantial volume. The principal impact of the "back door" imports seems to have been the creation of intense annoyance to other dealers facing customers claiming to have heard prices quoted up to 40 percent lower than the U.S. level.[19]

Given the dealers' sole and limited access to English imports, one is at first surprised to see so much of the saving being passed on to the farmer. This must

Table 10-4.   British Price Increases, May 1970–July 1971 (Prices in Pounds Sterling)

	M–F			Ford			IH	
	135	*165*	*1080*	*2000*	*4000*	*5000*	*434*	*634*
May 70	939	1,075	2,400[a]	822	1,109	1,224	882	1,281
Sept. 70	989[a]	1,190[a]	2,400[a]	919	1,211	1,399	1,007	1,391
Feb. 71	1,074[a]	1,290[a]	2,660[a]	1,053[a]	1,366[a]	1,650[a]	1,037	1,550
Apr. 71	1,074[a]	1,290[a]	2,660[a]	1,053[a]	1,366[a]	1,650[a]	1,307	1,550
July 71	1,164[a]	1,481[a]	2,941[a]	1,140[a]	1,518[a]	1,862[a]	1,037	1,785
$\frac{\text{July 71}}{\text{May 70}}$	1.24	1.38	1.23	1.39	1.37	1.52	1.17	1.39

[a]150 pounds have been deducted from declared list price to cover the cost of a fitted safety cab.
Source: *Agricultural Machinery Journal*, various issues.

be explained by the unusual nature of the transaction, by farmer doubts about whether he was getting exactly what he would if the machine purchased were of U.S. origin, and by the fact that the manufacturer's warranty was void; the farmer had to accept an ad hoc warranty from the dealer.[20] This suggests an obvious but important proposition: Where goods are durable and service is important, tying service to a specific dealer (or dealers within a certain area or country) may prove to be an effective means of limiting arbitrage when price discrimination is practiced.

Although the bootlegging of Massey-Ferguson, Ford, David Brown, British Leyland, and Deere tractors probably grew steadily in the mid- and late sixties, one estimate suggested that not more than 600 tractors entered the U.S. in that way during 1969.[21]

Farmers in the United States and Canada have faced similarly high prices for tractors relative to buyers elsewhere, with prices below the border usually higher than those above, but there has been an astonishing difference in the response of the agricultural communities in the two countries. While the differences in retail prices between the U.S. and Britain have gone almost unnoticed in the United States,[22] both private and official groups in Canada have discussed the subject widely at least since 1960, when public hearings were held.[23]

The Royal Commission on Farm Machinery, appointed in May of 1966, had as one of its principal assignments to investigate ". . . the cost to the user of agricultural machinery in Canada as compared with the costs of similar equipment to users in other countries. . . ."[24]

During the life of the Commission there was an extraordinary amount of activity at the provincial level, especially in Ontario. The Ontario Federation of Agriculture formally organized to buy tractors on behalf of farmers in all parts of Canada from agents in Britain.[25] The Ontario Special Committee on Farm

Income, an organ of the provincial legislature, attacked the high cost of machinery distribution through conventional channels by threatening to propose an Ontario Farm Machinery Crown Corporation for mandatory wholesaling so "farm machinery would be sold in much the same way as beer is distributed in Ontario through the brewers' retail system."[26] Manitoba also had a Farm Machinery Investigation Committee, and the Canadian Minister of Agriculture visited the Zetor tractor works in Czechoslovakia to encourage exports to Canada to put downward pressure on the price level there (but see p. 180).[27]

In March 1970, several months after the *Special Report on Prices* concluded that even after taking additional, and perhaps unwarranted, distributional expenses fully into account, at least a third of the price difference between the U.K. and Canadian machines was unaccounted for,[28] the *Canadian Farm Implement Dealer* editorialized: "How soon is it going to be before our 'big brother' to the South gets into the act? You know how hard some of 'his' senators kick when the get angry."[29] The magazine's alarm, although seemingly plausible, proved to have been unwarranted. With the exception of the manufacturer-dealer trade journal *Implement and Tractor,* no important U.S. publication carried accounts of the price differences documented by the RCFM or the intense Canadian activity aimed at fighting the relatively high prices.[30]

Perhaps even more than his Canadian counterpart, the British farmer can be characterized as being both militant and informed. It appears that most, if not all, county branches of the National Farmers' Union (NFU) have machinery committees, the purview of which is the price and quality of farm machines. The National Union has a corresponding section which has engaged in activities such as lobbying for parts standardization and designing "model" tractors.[31]

Although unrest about the cost-price squeeze is perpetual in British farming as elsewhere, in 1965 the NFU decided to instigate a machinery boycott as a protest against the government's meager subsidy offer in the annual commodity Price Review. Predictably, the tractor trade suggested that the wrong target had been chosen, and it is not clear that the boycott had any important impact. The problem arose again after the end of the economy's surveillance by the Prices and Incomes Board, however, and the subsequent explosion of tractor prices brought forth another announcement of official NFU boycott—one which held the potential for more far-reaching effects than the previous one.[32]

As Table 10-3 demonstrates, prices on some models rose up to 50 percent between May 1970 and July 1971. British farmers never seem to have been overly impressed by the industry's claim that they have over the years enjoyed the lowest level of tractor prices in the Western world, and the rapid rise of Zetor to a 5 percent share in two years (1966 to 1968), during which other tractor prices were rising only slightly, suggests the vulnerability of the British market to a rapid advance through price appeal. By September of 1971 county branches of the NFU were urging the government to remove the farm machinery tariff.[33]

A possible facilitating agent for the increased penetration of the British market by foreign machinery might be the Agricultural Committee on Trading, or ACT. Begun in 1962, as an off-shoot of the National Farmers' Union, ACT grew during the 1960s selling simple farm supplies but never had any luck getting a tractor or other large machinery manufacturer to sell equipment to it, not only because the makers would fear offending the ordinary channels of trade but because ACT never demonstrated an ability to organize after-sales service.[34]

On the Continent, agriculture usually aims its dissatisfaction with prices paid and prices received directly at the government, often in an emblematic and ineffectual way, although there have been from time to time major farm machinery boycotts in France and Germany.[35] The effectiveness of these boycotts cannot be discerned as they usually occurred during poor agricultural years when one would expect a drop in demand anyway.

As in the U.S. case, one looks in vain on the Continent for the kind of careful attention to machinery prices and alternative source of supply which characterize British and Canadian agriculture, and this helps explain why vast differences in wholesale and retail price levels among EEC members persisted after all internal barriers ended on July 1, 1968. Such machinery price publication as does exist is confined to minor national periodicals. Furthermore, Fiat, clearly the most conspicuous low-price, home-market seller within the original EEC, is not well known as a tractor-maker under its own name outside Italy. Its participation in Germany was negligible until the late sixties, and the large French share held by Someca was based on a product line containing enough French names and numbers as well as components to render arbitrage of "the same" product from Italy an uncertain enterprise. This argument, of course, only suggests why identical or very similar machines sold for different prices in different EEC countries as late as 1970. An individual German farmer, for example, trying to take advantage of cheap Italian equipment not sold in his own country, even if he knew it existed, would obviously face the same problem of parts and servicing as would his North American counterpart.

Although the level of Australian tractor prices has been substantially higher than in the principal supplying country, the U.K., throughout the period, very little attention by farm groups seems to have been directed at anything other than the avoidance of a tariff increase.

A major watershed in the development of Australian policy was Chamberlain's application to the Tariff Board in 1953 for a 37.5 percent duty on virtually all wheeled agricultural tractors; the firm's case was supported by the Chamber of Manufacturers. The Council of Agriculture, however, representing all the regional agricultural product associations, took a position which seems to have been sustained for many years thereafter: ". . . if it is found that the Australian industry requires some assistance, such assistance should be by continuance of bounty payments . . . [the burden of] development of secondary industry should be borne by the whole community rather than by one section."[36] Essentially the same position was taken by both sides a decade later, and the system remained unchanged.[37]

GROUP INCENTIVES TO GET TRACTOR PRICES LOWERED

Even if information about cheaper tractors were known to farm groups, it might not be obviously in their interests to take (presumably not costless) collective action to make such equipment available. If farm groups in countries with completely protected agriculture believed that all output prices would be immediately adjusted to yield farmers the same income as before the reduction in tractor prices, then the incentive for collective action would be clearly absent. This could conceivably approximate the situation in the EEC (*after* the unification of input prices which had not yet occurred by the early seventies), but, even there, the possibility would still exist for partial or temporary gains which would be worth some effort. A more general argument against collective action obtains where the government is not expected to engage in compensatory activity, but the lowering of tractor prices is expected to result in a non-negligible increase in commodity supply. Virtually all farm products are sold under conditions of inelastic demand in developed countries, and there exists the possibility that increased output could result in a lowering of commodity prices which would more than offset the increased income which the lower tractor prices would initially seem to promise. Empirical studies, however, have been able to find no substantial link between tractor prices and output; tractors are estimated to substitute almost entirely for labor when their relative prices change.[38] It is therefore not surprising that farm groups which have become interested in tractor prices have failed to take the possibly negative impact of lowering them into account. Even at a theoretical level, of course, the argument is irrelevant for individual users, national agricultural sectors facing elastic international demand, and virtually all groups of non-farm users.

CONCLUSION

This chapter has established a number of important facts about pricing in the world industry. First, there has been a substantial change in the pattern of internal prices among the markets in the study over time; the most notable change in this regard has been the secularly increasing relative price level in North America. Second, Italy has been established on the basis of internal prices to be a country of continuing low absolute cost advantage along with the United Kingdom. Third, at the end of the period, there was a marked tendency for North American price-cost margins to be higher in the power range beyond that of substantial foreign production. Several possible (and complementary) explanations were offered: the discouragement of real and potential foreign competition, different derived demand elasticities in different types of agriculture, and the penchant of affluent farmers for speed of operation. Fourth, widely differing international prices for similar or identical equipment have been sustained, and these differences are not explained by transportation costs or tariff barriers. Finally, there is remarkable international variation in farmer attention to foreign tractor prices; a partial explanation for this phenomenon will be offered in Chapter Thirteen.

Chapter Eleven

Overall Firm Strategies and Changing Sources of Supply

Despite the rather modest extent to which production extensions have occurred since the immediate post-war years, intrafirm sourcing patterns have nonetheless undergone a significant alteration, and these changes can be best illuminated within an attempt to discuss overall firm strategies and the relative competitive positions of the various firms. The established industry's behavior toward potential competition will also be examined, and the chapter will conclude with an overview of the development of production and trade in the world tractor industry.

MASSEY-FERGUSON

During the 1950s, Massey-Ferguson struggled with the problem of whether or not to retain separate dealerships for the previous Massey-Harris and Ferguson lines in its various markets and the related question of how different equipment sold through the separate dealerships should be. Once this issue was sorted out in favor of a uniform product and dealer line, management also concluded that the company's dependence on outside suppliers was excessive. This is an issue which has not been discussed previously; the degree of vertical integration which was presented in Chapter Three as typical of a North American tractor-maker applied to virtually all tractor-makers throughout except Massey-Ferguson and its two predecessor companies. Massey-Ferguson did not achieve the "normal" level of vertical integration until the late fifties; indeed, the manufacturing value-added by both firms before the merger and for several years thereafter was rather slight. Ferguson's North American machines and those of Massey-Harris relied heavily on Continental Motors for engines and on Borg-Warner for transmissions and axles; when these high-value components are added to those which are usually purchased by a tractor maker, anyway, little manufacturing remains. The value-added in manufacture in North America

171

by Massey-Ferguson on most of its tractors sold during the fifties may have been no more than 15 to 20 percent. In the U.K. the situation was even more extreme. The Standard Motor Company, Ferguson's original British supplier, was producing all of the firm's British tractors in the late fifties and supplying diesel engines for U.S.-produced machines. Massey-Ferguson achieved a vast increase in vertical integration in a very short time by purchasing Borg-Warner's Detroit transmission and axle plant in 1957 for a little over 20 percent of its reproduction cost, and, on the other side of the Atlantic, by purchasing Standard's tractor-making facilities (though none for engine production) in 1958 and Perkins Diesel in 1959. The firm continued to buy gasoline engines from Continental and General Motors for some of its North American units during most of the sixties.[1]

Massey-Ferguson's management had realized, even before its vertical integration moves, that great economies could be realized from a common international design and centralized manufacture of major parts. By 1960 this had been accomplished for nearly all of the firm's tractor production outside of Continental Europe. On the Continent, a complex of contractural sourcing arrangements made common design more difficult, but with the introduction of the 1965 line, complete interchangeability among machines produced in France, the U.K., and the U.S. had been achieved. An expansion of Massey-Ferguson's French facility in 1964 seems to have been simply a matter of minimizing the firm's delivered costs. Britain did not appear to be joining the Common Market, and so it was cheaper to manufacture some components in France; there were also plans for exporting some French components to the U.K. and the United States.[2]

The least readily explicable aspect of Massey-Ferguson's behavior is what appears to be a continued uneconomic reliance on North American fabrication, but this is probably due at least in part to the firm's marketing strategy. Although nearly every machine sold in the U.S. and Canada in the late sixties had a British engine, its rear axle and transmission, some stampings, machined parts and castings, as well as many outside purchased components were usually sourced from North America. In fact, while foreign content for smaller tractors was greater than for larger, approximately half of the value of all tractors sold in North America in 1970 was claimed by the company to be of North American origin.[3] Why? Massey-Ferguson has apparently always believed, rightly or wrongly, that U.S. farmers are resistant to foreign machinery per se (but see above pp. 87-88), and the firm may have believed itself to be doubly vulnerable if it were seen as a Canadian firm which simply assembled British components. North American headquarters were moved from Toronto to Des Moines in 1965 expressly for marketing purposes with the objective of persuading ". . . mid-West United States farmers to regard Massey-Ferguson as a company permanently committed to the United States market."[4]

Despite its bargain price on the Borg-Warner facilities, Massey-Ferguson has never denied that British production was far cheaper than North American, and in 1970, during the height of the furor over the *Special Report on Prices*, a few small 135's were imported from the U.K. into Canada (*not* into the U.S.) for sale at lower prices than Detroit-sourced units, with no suggestion that they were in any way inferior.[5] Further considerations concerning the continuation of North American manufacture will be presented in the section below dealing with International Harvester's behavior.

FORD

Like Massey-Ferguson, Ford by the mid-sixties had a worldwide line of virtually identical tractors, with even less duplication of parts manufacture in various locations, but the developments which lay behind this outcome were quite different. After the breach with Ferguson, Ford (U.S.) continued the manufacture of a machine very similar to its previous model. The British Ford company, however, went its own way, finally imitating the essentially unchanged Ferguson machine in 1952 with its more powerful "Major" which was complemented by the smaller "Dexta" in 1958. These diesel machines were imported by Ford (U.S.) into North America as whole tractors and sold along with Detroit's gasoline models. By 1960 the imports may have accounted for as much as 25 percent of all Ford's North American sales,[6] but there was no attempt to coordinate either technical design or physical appearance between the U.K. and the U.S. Ford attained full control of its British subsidiary in 1960, however, with the declared overall objective "to obtain greater operational flexibility and to coordinate better its American and European operations."[7] The main reason for the move was the freedom it gave the firm in automobile design and production, but it was followed a year later by the setting up of a separate division to handle all tractor design and manufacture worldwide. A new line was introduced in 1965 which utilized a new and larger factory at Basildon, near London, the output of which was to be complemented by assembly operations in Detroit and in Antwerp, where an erstwhile Ford car assembly plant had just been refurbished for the purpose. There was then virtually no production overlap except for a few stampings. The most sophisticated transmissions were produced only in Detroit, while certain other transmissions and rear drives were produced in Antwerp. Most other parts came from Basildon.[8] Even the very largest tractors introduced only into North America in the late sixties used British engines, although most of their other parts were locally sourced.

Although Ford's declared strategy, to maintain or extend its non-North American role, while devoting primary attention to a very substantial increase in North American share, is similar to that of Massey-Ferguson, it apparently was little concerned about the importance of U.S. content for market penetration. While U.S. assembly may be in part for marketing purposes, it amounts to only a few percent of cost, and the cost of the policy is reduced by the fact that the U.S. duty (11 percent before January 1, 1972, and 5.5 percent afterward) on industrial tractors, one-third of Ford's sales, is thereby avoided.[9] It is further possible that the Detroit operation may be diminishing in importance. While the 1965 line of tractors was supplied to Canada from Detroit assembly prior to 1967, direct importation from Britain for all but the largest models was then begun.[10]

In the past where Ford and Massey-Ferguson's differences in tractor strategy differed most markedly was with regard to the Continent. Ford made no attempt to develop machines suitable for the bulk of Continental use during the fifties, as the U.K. group concentrated on medium-sized tractors which were in heavy demand in Britain, North America, Australia, and Scandinavia. Ford's Continental distribution remained rudimentary, despite the fact that in their power range, Ford machines could easily match the prices of similar equipment in most European markets, despite tariff barriers.

Ford's reluctance to develop a range of different models must be partly explained by the success it had with the machine it developed in the early fifties. The sterling area and later European Free Trade Area markets were almost invariably jointly dominated with Massey-Ferguson, and both presumably saw the profit reduction from increasing the number of models offered. Massey-Ferguson, of course, did develop additional small models especially geared to Continental demand (and produced at high Continental cost), but Ford's management on both sides of the Atlantic was unwilling to devote substantial managerial talent or capital expenditure to tractor manufacture, an endeavor treated very much as a corporate afterthought. Industry lore suggests that it was Henry Ford II himself who noted the vast and growing profit potential of farm machinery and who was directly responsible for the consolidation of all tractor activities under one management division in 1961. The establishment of the Antwerp plant in 1964 and a reorganization of European distribution in mid-sixties to increase contact with retailers—especially in Germany—were a recognition of the importance of the larger unit Continental market and the extent to which it could be served by essentially the same tractor line marketed elsewhere. Its declared intention to broaden its European offerings beyond tractors and some complementary implements was yet to be realized in 1970. [11]

INTERNATIONAL HARVESTER

Although Massey-Harris had been active in farm implements in many overseas markets before the war, both through exports and domestic fabrication, only International Harvester of the North American producers was an important exporter of tractors. As outlined in the entry chapter, the firm took swift action in France, Germany, Britain, and Australia not only to maintain its tractor position but, far more importantly, to share in the post-war boom.

These moves, however, were handled in a piecemeal fashion which resulted in International Harvester's lagging considerably beind Massey-Ferguson and Ford in developing a common line of tractors with components designed to be produced at maximum volume. Throughout the entire period, International Harvester's non-U.S. activities were handled by an Overseas Division, with French, German, and British operations under one general manager, and Australian and other far eastern ventures under another. Furthermore, during the fifties, not even the European subsidiaries had much contact with each other, although some diesel engines were provided to the French by the German subsidiary in 1959 in return for castings.[12]

In 1961, in recognition of the forthcoming Common Market, a European Planning Office was set up to coordinate the production of engines and hydraulics in Germany and transmissions and axles in France for a common Continental line of tractors which appeared in 1966 and was assembled in both countries. The British factory was partially integrated into the scheme, but the entire project appears to have taken place independently of North American tractor planning which remained lodged in the U.S. Farm Equipment Division.[13]

The Australian subsidiary functioned throughout as an assembler of imports from Britain and the U.S., while also producing some of its own components. In the mid-sixties one model was advertised as the only machine in the country with an "Australian-manufactured engine,"[14] presumably one obtained from the company's substantial truck manufacturing facility in that country.

The idea that foreign production might do more than simply replace erstwhile imports seems to have been thrust upon International Harvester by Ford's importation into North America of the "Major" in 1954. Five years after Ford's first British diesels came to North America, International Harvester began importing a small diesel from its British subsidiary as an expressed reaction.[15] Small diesels from Britain remained in the North American product line throughout the sixties, and a few thousand machines were imported each year.

While marketing advantages, real or imagined, may play some role in keeping Massey-Ferguson's North American fabrication higher than it would otherwise be, why would International Harvester have avoided an expansion of its European, particularly its British, production to serve the North American market, even if some U.S. assembly were retained, as in the Ford case? (International Harvester's president predicted a shift toward greater reliance on non-U.S. sources of supply based on cost advantages in 1962.)[16] Of course, Ford and Massey-Ferguson's ability to operate high-volume, low-cost operations in Britain did not prove that International Harvester could. Nevertheless, International Harvester admitted (see pp. 42) that production costs were much lower in Britain in 1960, despite its modest scale operation, and the direction for the firm to take would seem to have been to expand its British operations at the expense of those in the U.S., even though this would undoubtedly have necessitated some internal reorganization of the company. Why did International Harvester not do so? In the fifties it is possible that British manufacture did not appear clearly cheaper to Harvester because of sunk capital costs in the U.S. (The relevance of this consideration will be more fully explored in the following chapter.) However, no completely new tractor plant was built in the U.S. between 1950 and 1970,[17] and the age of tractor production machinery (and there is no evidence that International Harvester's is any exception) came to be among the oldest in all of U.S. manufacturing (see Chapter Twelve). Thus, by the end of the period, this could certainly not be the whole story, and one must look further for an explanation.

While shifting the manufacture of a product to a foreign country involves some mobility in management ranks, many new faces emerge. There would undoubtedly be considerable unrest among middle management at the prospect of either moving abroad or finding new employment, and over the years International Harvester's top management has been regarded as one of the most complacent and least profit-hungry in American industry,[18] a factor which helps explain its poor tractor (and other farm machinery) performance in North America. Of still greater importance would be the replacement of the labor force. Because International Harvester was not a rapidly growing firm, any substantial shift to foreign sources of supply would lead to a major reduction in U.S. employment. As might be expected, the United Automobile Workers (UAW), the union representing tractor workers, has well-developed views on foreign fabrication. Foreign plants must not, in the words of UAW Vice-President Ken Bannon, ". . . have the effect of transferring, directly or indirectly to Britain or Germany work now being performed by UAW members in plants in the U.S."[19] Heated controversy on the subject was aroused even when foreign manufacture of some components was persuasively argued by the automakers to be the only means of staying increased penetration of the U.S. market by foreign companies. How indefensible would such a move seem if it were proposed only to increase profits? Even Ford and Massey-Ferguson, let alone truly foreign firms, had not dramatically increased their North American market share by the late sixties. This suggests a general hypothesis: Where labor is well organized, a substantial shift to lower cost sources of supply may be extremely difficult if not impossible when simply to increase profits. Only if the workers displaced can be absorbed into the home production of other products, or under the most dire competitive threat from abroad, will the extension of foreign operations at the expense of those at home be observed.

DEERE AND COMPANY

The last of the large North American firms to gain foreign experience was Deere and Company, the melancholy Continental adventures of which have already been related. Deere, unlike the three firms just discussed, did not have substantial export markets from the pre-war period to protect and extend, and its acquisition of Lanz in 1956 was its first foreign experience in a major market. Why did Deere go abroad? Three motives were expressed: shrinking export markets, new and growing markets abroad, and the threat of foreign competition based on economies of scale.[20] The first concern seems principally to have motivated Deere's willingness to engage in various development projects, particularly in Latin America, where early post-war demand had been substantial, but the other two seem directly related to its Lanz venture. Deere was confident of its marketing skills, for at the time of the Lanz acquisition, its marketing strength alone was pulling it ahead of International Harvester in the U.S. and Canada, and U.S. management saw European distribution as ripe for reorganization along North American lines.

The ultimate domestic impact of not going abroad was also considered. Deere viewed with alarm the attempts of Massey-Ferguson and Ford to expand in the North American market on the basis of volume gained elsewhere and compared the open U.S. market to a region within the U.S.:

> . . . if we were to decide not to compete in the entire free world market, it would be almost as if we had decided years ago we would manufacture and sell our produce in Iowa, in Illinois—instead of being a national producer. In time, manufacturers selling to a bigger market, being able to spread their cost of such items as research and development over a larger volume, would be able to run us into the ground with their competition.[21]

Although economies of scale in research and development were probably given only as an example and not because Deere regarded them as of primary importance, one does search in vain for an explanation based on underlying factor costs. As pointed out in Chapter Nine, an essentially American-designed line was introduced in Germany in 1960, and a new line of engines with interchangeable parts was introduced in both Dubuque and Orléans in 1965,[22] but there was no significant shift to European sourcing.

Deere was very slow to introduce European equipment into the United States, perhaps fearing a negative reaction to foreign fabrication. It sold two Mannheim models in Canada during the sixties,[23] and this may well have been an attempt to test the North American waters. The first German import into the U.S. was the bottom-of-the-line "820" introduced in 1970.

It is entirely possible that Deere has had little cost incentive to expand European operations at the expense of those at home. Its Mannheim plant, despite some revamping, was never modern, and given the level of variable input costs, particularly that of skilled labor, a really substantial renovation would have been risky.

The probably unimpressive level of German absolute cost and the small volume of production at Mannheim must have contributed considerably to the attractiveness of the proposed combination with Fiat in 1970 noted in Chapter Eight. Not only did it offer the prospect of equipment production at lower cost but it probably appeared to blunt the development of international marketing potential by Fiat; such marketing expertise could possibly someday be directed at the North American market. For Fiat, the venture offered Deere's hard-won foot-in-the-door in several Continental markets. On the other hand, a joint venture between large firms covering a vast number of products and markets is an enormous commitment with innumerable areas for possible disagreement. The different approaches of the two firms in marketing alone may have been an insurmountable stumbling block. Furthermore, Fiat was worried about the possible political impact of any production consolidation and the implied reduction of employment in some areas.[24]

OTHER FIRMS

None of the other firms seems to be pursuing nearly as grand a strategy for tractors as are the "big four." Fiat's lack of marketing knowledge is undoubtedly a crippling factor, and this in turn must stem in large part from its home market policies. While the North American-based giants were experimenting with different methods of wholesale and retail distribution in the fifties and sixties, Fiat disposed of most of its tractor output by doing little but opening the factory door. Despite its almost certain advantages over other Continental firms in both scale and labor costs, it had made a good foreign showing only through Someca in France up to 1970. Even the potentially lucrative German market was almost completely ignored until the last few years of the sixties.

A further indication of Fiat's marketing weakness is its willingness to give apparently unrestricted licensing for the Rumanian manufacture of some models.[25] The company has given no official explanation for such generous licensing, which allows the Universal to be sold where Fiats might sell instead. Similar arrangements for automobiles have been justified with the argument that if Fiat had not offered such generous terms, a competitor would have and that "each time a car is sold it raises overall Fiat penetration. Above all it is a sale that does not fall to Western competitors."[26] Such arrangements loom large in total Fiat activity; over a half million of Fiat's 2.3 million 1971 output of vehicles was done under foreign licenses and by affiliated companies.[27] Whatever the efficacy of the automobile strategy, this writer finds it difficult to believe that any of the major companies in farm equipment would have made such a generous arrangement with the Rumanians, and it does seem rather odd to find two almost identical machines selling in France in 1970, with the Fiat selling at a 20 percent premium over the Universal.[28] The lack of an implement line to complement its tractors is another important Fiat weakness.

The retention of substantial French content on some Someca models is somewhat puzzling, but Fiat's tractor demand grew so rapidly in the sixties that French capacity might have been welcomed at least temporarily, while cutting back on French activity could create serious labor difficulties, of which the firm had more than its share at home. The practice also facilitated price discrimination.

Renault's market strategy at the end of the sixties was apparently directed mainly at speeding the pace of product development, organizing a full line, and improving distribution in the home market. Previous attempts to become an important power in Germany and Italy and earlier in Australia and North America had come to very little. Although there were reports in the late sixties that the farm equipment division of Renault was under pressure to improve profitability,[29] a source close to the organization noted that this eventuality is unlikely. Management is not aggressive, the management team is small, and most are technical people. The source points out that, given Renault's role as a state firm and the number of employees involved, the tractor operation cannot either be sold or shut down—whatever its financial fortunes.[30] This seems to be confirmed by a statement made by Renault's export director at the time of the introduction of the 1965 tractor line: "Of every 10 tractors made in France, eight are from firms of foreign origin, and, as France puts much store in her agriculture, an independent national enterprise is indispensible."[31]

Deutz's overall strategy is difficult to discern. Unlike Renault (and like Fiat) this leading national manufacturer has not seen its share of the home market eroded by the large North American firms. Nevertheless, it has managed to establish itself with a substantial share in only one major foreign market: France. Attempts in the late sixties to become established in North America seemed unpromising because the firm would not, or could not, attempt to compete on price. In the home market, Deutz, like Renault, is defending its position by building a fuller line which may increase its appeal abroad as well.

David Brown appeared to be maintaining its competitive strength both at home and abroad up to 1970, despite small volume, because of its modern facilities, low labor costs, expertise in geared drives and hydraulics, and the reputation for high quality which remained unsullied over the entire post-war period. Nevertheless, the lack of a broad range of tractors and a full-line seriously limited its ability to expand in North America during the sixties, and, by the end of the decade, the latter disadvantage may have been limiting its appeal on the Continent as well. Since the material for this study was prepared, David Brown's tractor operations were sold to J.I. Case, and shortly thereafter Case ceased to make farm equipment other than tractors and tillage tools, apparently hoping that a very long and dense tractor line and an array of construction equipment would provide enough appeal to retain dealers.[32]

British Leyland's position in the late sixties strongly paralleled that of David Brown in several respects. They both produced from new plants at approximately the same volume (although Leyland's volume in engine production—much of which was relevant—was of a wholly different order of magnitude), and both sold a narrow range of tractors in most markets at a significant discount from their major competitors. David Brown appeared to be pursuing the latter strategy more consistently in the major markets, however, while British Leyland seems to have been giving more attention to the "periphery" by showing a willingness to engage in various kinds of joint-ventures in developing countries in which the company was at the same time attempting to build distribution systems for cars, trucks, and buses.[33]

Ever since the late fifties, the demise of the small U.S. full-line producers has been predicted, but the combination of high margins on large tractors (and on combines) and sourcing of smaller tractors from abroad has kept them afloat. The future of these firms clearly depends on whether the margins on more powerful tractors can be sustained and/or the firms can establish lower cost sources of supply. With the 1972 purchase of David Brown, Case appears to have vastly increased it competitive position in tractors and simultaneously removed a source of price competition which could well have progressed into higher horsepower ranges over time, but it soon gave evidence of its inability to produce other complex farm machinery profitably. White Motor purchased a combine factory in Italy in 1966, and, although it was not immediately used to source combines for North America, the increased sourcing flexibility must have been one reason for the move. In addition to selling some Fiat machines in North America, White moved to blunt the thrust of another possible future competitor in the North American market by reaching an agreement in 1969 with the small, but rapidly growing, Same. White arranged to market all of Same's tractors outside of Italy and North America (although none was sold in North America); whatever international marketing experience Same might have gained was thus largely lost.[34] What other firms remain? The high-cost, low-volume Volvo appears to have little future outside of Sweden, although its entrenched name acceptance and distribution system may well allow continued but hardly expanding participation in the home market. The North American specialist firms such as Versatile (and their counterparts elsewhere) could be driven to the wall at the whim of the major firms and will continue participation only under their umbrella.

Ever since the war's end, the Czech Zetor has been sold in small numbers throughout Europe, and in the mid-sixties a determined and successful attempt to expand market share was begun in Britain, on the Continent, and in Australia. Zetors were not sold in North America, however, and the RCFM found that Deere negotiated for the North American distribution rights for the Zetor tractor during 1970 [35]—sources close to the industry report that the reason that Deere sought exclusive North American rights was simply to keep Zetor out of the market; the outcome of the negotiations is not known, but Zetor did not enter.

The Rumanian Universal appeared in the early seventies to be the most likely source of sustained North American competition from Eastern Europe when it began to be distributed in 1972 in the U.S. by one of the same distributors that handled British Leyland,[36] but, while the Eastern European role may expand somewhat, sales resistance may continue to prevent the achievement of a large market share in North America unless very strong antipathy develops toward traditional sources.

Japan has often been an important source of world competition in manufactured goods, but Japanese experience with agricultural engineering has been very limited. One Japanese firm entered into a licensing agreement with Ford as early as 1950, and during the sixties licensing agreements were reached between various Japanese firms and International Harvester, Perkins, and Deere.[37] When approached by the RCFM in 1970, Japanese trade representatives in Ottawa reported simply "that Japanese firms have entered into technical and licensing agreements with major tractor manufacturers on this continent, which would prevent them from manufacturing tractors for sale in either the Canadian or U.S. markets."[38] It is possible that non-entry could have been an important part of the agreements, and, whatever the value of the technology involved, it would then be sold at a discount corresponding to the comprehensiveness of the non-entry arrangements and the desirability of entry by the licensee in the absence of the pact. In tractor manufacture, the technology itself would not appear to be of much value, at least by the late sixties, as is shown by the growth of the imports into North America of small Japanese machines of wholly satisfactory design from unlicensed firms. Furthermore, as Caves has pointed out, today's licensee may very well be tomorrow's competitor [39] (unless he can continually be "bought off"). Nevertheless, from the perspective of the late sixties, any policy of buying time vis-à-vis the Japanese might have appeared attractive, because the growth rates projected for Japan implied a rapid diminution of its absolute cost advantage.

RELATIVE COMPETITIVE STRENGTHS

The discussion of the previous chapters has led us to identify some key characteristics of competitive strength: a broad line of equipment sold through closely supervised dealers, the production of tractors in areas of low absolute cost, and large production volume. If these three considerations are crucial determinants of long-run strategic advantage, then it is quite clear why no firms other than the "big four" and Fiat faced a promising future in the early seventies. Nearly all other firms lacked at least two of the three attributes. Furthermore, if all three prove to be important for long-run viability, Massey-Ferguson stood alone in worldwide strategic strength in the early seventies. In addition to established experience with full-line selling and close dealer contact in most markets, Massey-Ferguson's production facilities were such as to allow it to profitably match prices almost anywhere in the world. Ford and Fiat's cost advantages as late as the early seventies were not balanced by product line and distributional strength (although in the Ford case the situation appeared to be improving rapidly), and International Harvester and Deere were high-cost producers.

Deere will probably continue to enjoy undisturbed prosperity and leadership in North America for many years unless Ford and Massey-Ferguson begin to use their cost advantage to put substantial downward pressure on prices. Were this to be done, Deere might well be forced to find a lower absolute cost source for much of its equipment. There can be little doubt, however, that it would have a considerable time to do so, because its customers are probably the most loyal in the industry. If Deere did seriously seek such a low-cost source, it would almost certainly not be Germany. International Harvester, too, could be forced to find a lower cost source of supply in the event of increased price competition, and its customers have historically been far more shiftable than Deere's. On the other hand, International Harvester has the apparent option of expanding its British operation if competitive conditions make that necessary. Both Ford and Massey-Ferguson seem to be well placed to continue to expand their North American share, and there appears to be nothing to prevent them from doing so; the real question is how much they will use price competition rather than gradual distributional development in their strategy.

Outside of North America, Ford and Massey-Ferguson are holding their own or increasing their role in most markets, and if Ford develops a full line, its continued success will be assured. International Harvester, too, may well at least hold on to its non-North American market share, especially in light of the coordination of European operations, but overall it appears to have no strengths which are not exceeded by Massey-Ferguson. Nothing in Deere's record appears to promise substantial improvement in its overseas performance; it faces an uphill battle, and it seems to have seriously overestimated the power of North American marketing practices in increasing market share. Fiat's potential in competing with the other large firms outside its home market based on its cost advantage is very real but is yet largely undeveloped.

RESEARCH AND DEVELOPMENT

The experience of the industry leaders confirms the oft-made observation that research and development are kept close to the firm's main headquarters because of scale economies.[40] Deere, Ford (after 1961), and Massey-Ferguson have done only a modest amount of design and prototype testing outside of North America. This is to be expected for the industry, wherever the bulk of production takes place, because new machines are usually large, and their first sales are overwhelmingly in North America.[41]

International Harvester has been the exception in allowing separate design by its manufacturing subsidiaries, but this appears to have resulted more from organizational inertia than conscious policy, and it diminished in the sixties, although all research and design did not become centralized in the U.S.

PRODUCTION AND TRADE IN THE
TRACTOR INDUSTRY

Tractors and the "Product Cycle"

One way of viewing the development of the tractor industry over time is provided by the Vernon "product cycle" model.[42] Relatively high labor costs in the United States triggered the invention and refinement of a producer good, the tractor, which was subsequently exported in substantial numbers to those countries with similar demand conditions (Canada, Australia, and certain parts of Latin America).[43] During the pre-war period most countries in Europe had farm sizes too small and labor costs too low to be important as export markets. Later, however, rising labor costs and increasing farm sizes in Europe (and, in part, government policy) created demand for the largely standardized product, and European production was begun, some under the aegis of the pioneering firms (often under threat of losing their export markets) and some through imitation by local producers.

The history of the industry up to this point fits the Vernon model well. What is hypothesized next to transpire, however, is that long-run comparative cost conditions should begin to tell and that the original high-income innovating market should be threatened by the worldwide diffusion of the innovation and multiple sources of supply. What much of the preceding discussion has demonstrated, however, is that the product differentiation barriers to entry enjoyed by the major North American (and to a lesser extent German) producers in their home market has to a large extent retained for the North American industry what international diffusion of an innovation through the product cycle is supposed to erode: rather low elasticity of demand for the domestic producers as a group. This, in turn, has given each major firm great discretion in its sourcing policy and has retarded a shift to lower cost sources of supply.

Figure 11-1 illustrates the overall production and foreign trade experience of the United States with the agricultural tractor. Two export series are shown, one in which Canada is counted as an export market and the other in which it is netted out. The U.S. industry saw two important periods of substantial sales outside of North America: one during the initial development of the tractor in the twenties and the other in the immediate post-war period when countries within the dollar area (mainly Latin America) added briefly to burgeoning home demand; these markets were subsequently lost to tractors imported from other countries (mainly Britain) and to development schemes (in which the major U.S. firms sometimes participated). Exports had greatly receded by the early 1960s when they amounted to only about 10,000 units excluding Canada. The subsequent slight rise appears almost entirely attributable to tied-aid financing by the Agency for International Development.[44]

If one looks only at whole machine imports as shown in the figure, the impression of "fortress America" in unavoidable. Not only did imports remain modest, but the lion's share of them came in under the aegis of the large es-

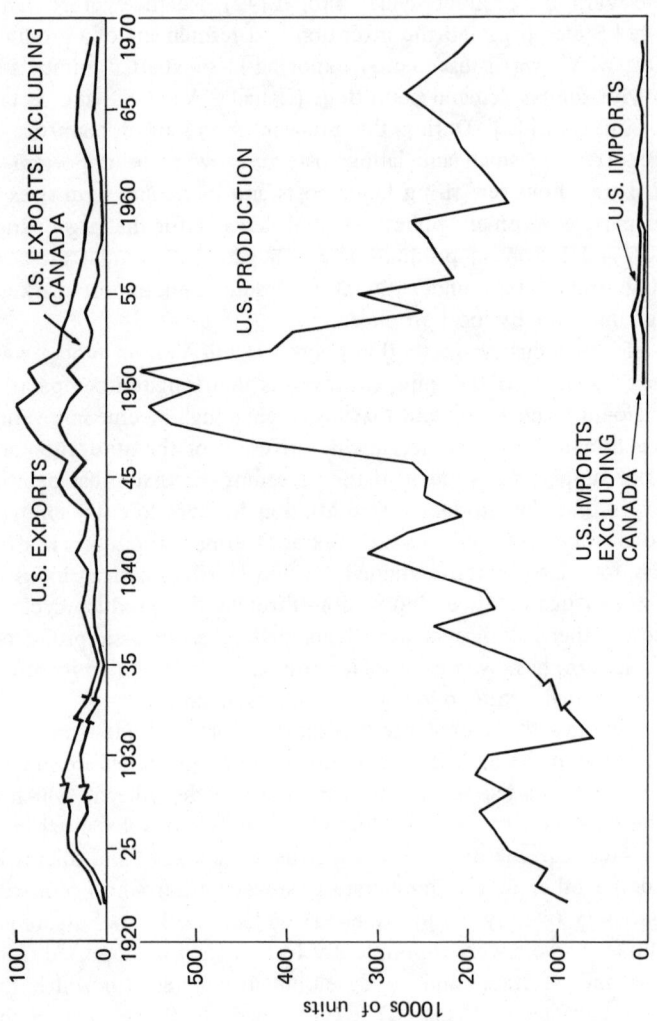

Figure 11-1. U.S. Production and Trade in Wheeled Tractors

Source: Production and Exports: Bureau of the Census, *Current Industrial Reports*, Series 35S, various years.

Imports: Bureau of the Census, *U.S. Imports*, Series FT 135, various years.

tablished firms. Confining attention to the importation of whole machines is, of course, quite misleading. In 1966, for example, if one takes Massey-Ferguson's own estimate of its non-North American content of machines sold there as one half, and makes the not unreasonable assumption that Ford's non-North American content was probably more like three-quarters, the "whole machine equivalent" imports for the United States in that year was an additional 40,000 units or 18 percent of the U.S. domestic market.

Non-North American Tractor Production

Previous discussion of national units outside of North America has stressed their role as tractor buyers and their relative vulnerability to imported machinery. Figure 11-2 puts the non-North American markets into perspective as locations of production.[45] Australia is not included in the figure; its miniscule "hothouse" production has already been discussed at length, and, not surprisingly, few machines have ever been exported. The handful that have were sold almost entirely to nearby pacific markets.[46]

The U.K. export role has been so important that it has already been discussed in previous chapters. The U.K.'s early start by comparison with other European producers was followed by a penetration of non-dollar export markets, aided greatly by the 1949 devaluation of the pound. The lead in European production was held in nearly every year, and Britain temporarily surpassed the United States for the first time in 1960. From 1948 onward British production for export exceeded home absorption, and almost immediately exports began to account for 70 to 80 percent of output, a figure which was sustained throughout. Furthermore, as was stressed in Chapter Eight, the export role of the smaller British producers (with the possible exception of Allis-Chalmers) differed little in relation to their domestic sales from that of Massey-Ferguson and Ford.

Both French and German production, which were far more controlled by domestic demand than that of the U.K., peaked and fell during the fifties. French exports prior to the massive devaluations of the late fifties were only a few thousand units of which over half went to Algeria alone.[47] The devaluations and the staged tariff reductions in the EEC subsequently allowed exports to increase considerably. By contrast with France, German output was exported in substantial quantities almost from the very beginning of large-scale production, with most of it going to nearby European countries; exports were 25 percent of production in 1955 and were sustained at that level. Although exports by firms are not available, there is much evidence that Massey-Ferguson and Renault in France and Deutz in Germany have accounted for export sales throughout vastly out of proportion to their shares in the producing countries.[48] Furthermore, as Germany's relative price level increased during the sixties, Deutz attempted to protect its share in several European markets by increasing price discrimination against home market customers.[49]

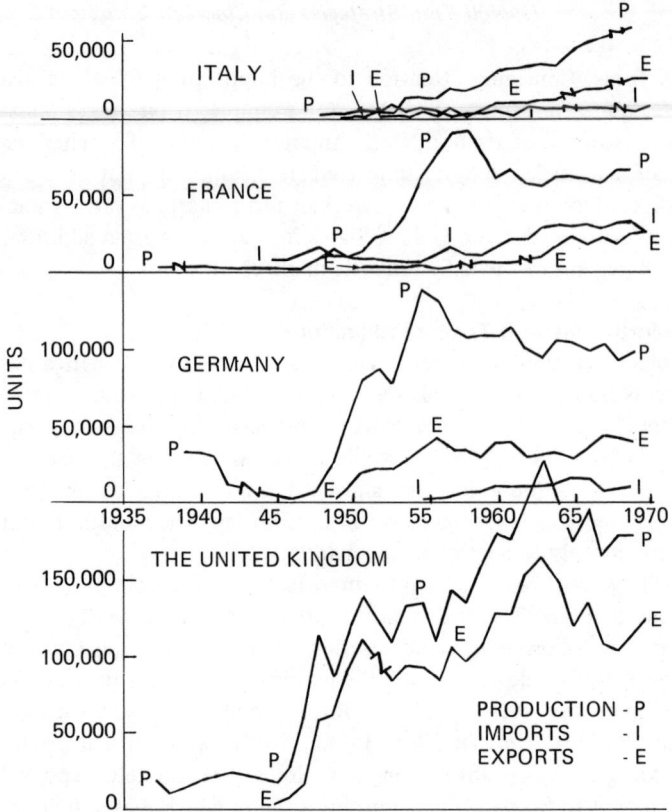

Figure 11-2. Tractor Production and Trade, Selected Markets

Source:

Italy: *Production:* U.N., *European Tractor Industry,* p. 37; *La Meccanizzazione Agricola in Italia; AMJ,* May, 1969, p. 49.

Exports: Farm Implement and Machinery Review, August 1, 1951, p. 636; Ministero del Commercio con l'Estero, *Statistica Annuale del Commercio con l'Estero,* various years. In foreign trade figures until 1962, no distinction is made among wheeled, crawler, and truck tractors. The 1963 figures lump wheeled and crawler machines together. In 1964, approximately 80 percent of the broad "tractor" category was composed of standard wheeled tractors. This percentage is used to estimate exports for the period of rapid export expansion back through 1960. Before that the gross figures are used; truck tractors are believed to have been a small number throughout so the principal error is the inclusion of crawlers.

Imports: For the period prior to 1964, imports are estimated

Figure 11-2 shows the rapid rise of Italian production after the mid-fifties and the even more rapid rise of Italian exports. Italian exports have increased in nearly every year since the early fifties, yet the really big jump in external performance came in 1960 when exports increased by 50 percent over the previous year. This year saw a comprehensive push, dominated by Fiat, which won gains in several European markets but also in virtually every other market

---

from registration figures; 98 percent of tractor imports from 1965–70 were wheeled.

France: *Production:* Syndicat, *L'Industrie des Tracteurs,* pp. 10, 21. (Figures from 1941 to 1944 include trucklayers); C.N.E.E. M.A., *Bulletin d'Information,* various issues.

Exports: Direction Générale des Douanes et Droits Indirects *Statistiques du Commerce Extérieur.* The 1951–1955 figures include a small number of truck tractors. These figures do not reflect a substantial export of tractor parts. In 1970 while whole tractor production was 68,800, the value-equivalent of parts production was equal to an additional 14,200 units. C.N.E.E.M.A., *Bulletin d'Information,* May, June 1971, p. 55.

Sales of Imports: Syndicat, *L'Industrie des Tracteurs;* C.N. E.E.M.A., *Bulletin d'Information,* various issues.

Germany: *Production: Landtechnik,* October, 1961, p. 658; U.N., *European Tractor Industry,* p. 34; *FIMR,* June 1, 1966, p. 657; *Landmaschinen Markt,* various issues.

Exports and Imports: Landtechnik, October, 1961, p. 651, 658; Statistisches Bundesamt, *Der Aussenhandel der Bundesrepublik Deutschland,* various years.

The U.K.:   *Production:* (since January 1, 1964, deliveries): Data from Ministry of Agriculture, Fisheries, and Food and Board of Trade, reproduced in *AMJ,* various issues.

Exports: Overseas Trade Accounts of the United Kingdom (London: Her Majesty's Stationery Office), various years. Figures do not include machines of less than 10 h.p. and may include a small number of crawlers.

Imports: There is no way of making a reliable estimate of wheeled tractors imported into Britain between 1953 and 1969 because only weight and value data are available and for a broader category than wheeled agricultural tractors until 1961, after which unit figures still included crawlers until 1970 when 168 were imported.

in the world. Second-place Same shared Fiat's export boom in the sixties,[50] a fact almost certainly linked to White Motor's successful attempt to take over the firm's non-Italian marketing.

By the late sixties French and Italian exports were of about the same order of magnitude, with the German level about one-third higher and British exports between three and four times that of its EEC rivals. There was a marked difference in the composition of the exports among the four. French and German exports were destined mainly for other Western European markets, with each of the two countries taking about 30 percent of the other's exports. French exports to other EEC countries alone were over 60 percent of its total and German exports over 50 percent. In contrast, Italy sold less than a quarter of its exports to its EEC partners, and the major part of its exports went outside of Western Europe. The U.K. exported under 10 percent of its total to the six, while the EFTA countries absorbed only another 15 percent of the total. What these figures imply is that production from countries with the lowest cost sources of supply dominated those non-EEC markets without national industries of their own. Furthermore, they illustrate that Fiat, accounting for as much as three-quarters of Italy's exports,[51] has met with greatest success in minor markets not included in the present study.

Linder's Hypotheses

Nothing in Chapter Three suggested that producing only a handful of tractors in a particular power class would not add significantly to costs, especially if the machines were quite beyond the size range being produced in greatest volume. Thus, the observation by Staffan Burenstam Linder that, under conditions of decreasing costs, countries will export goods most heavily favored in the domestic economy and import goods demanded by minorities, is of some explanatory power in the tractor industry.[52] The handful of the very largest American units sold in Europe represents one example of such a phenomenon, as does the importation of small diesel units into the North American market during the fifties and sixties from European subsidiaries or independent firms. In the case of North America, of course, the confirmation of the hypothesis adds nothing to an understanding of why importation remains so minor given the underlying absolute cost differences.

Another of Linder's predictions, that free trade will increase intra-industry international penetration most for countries at similar levels of per capita income,[53] appears to be confirmed by the increase in tractor trade between France and Germany with the inception of the EEC because the ranges of models offered were for approximately the same type of agriculture and comparable costs for other agricultural inputs.

Chapter Twelve

Performance

The performance of the world tractor industry should be measured against a number of criteria: profitability, efficiency of production within a single country, international production adjustment, progressiveness in manufacturing technique, product development and quality, and efficiency in distribution.

PROFITS

As suggested in Chapter Three, RCFM estimates of economies of scale in North American tractor manufacture combined with a representative transfer price for an "average" tractor in the late sixties produced very high rates of return on manufacturing assets at the higher volumes of production considered. The after-tax internal rate of return was only 6 percent for a 20,000 unit plant, but it was 16.5 percent for the 60,000 annual output and 22.5 percent for the 90,000 unit operation. Estimates of British costs were at least 25 percent lower, thus generating extremely high profits at American prices at all three volumes. A look at actual company behavior revealed that half the content of Massey-Ferguson's offerings in North America (by value) were of British origin by the mid-sixties, that Ford's percentage was substantially higher, and that the U.K. components of these two firms were produced at volumes exceeding any studied by the RCFM. Deere and International Harvester had not engaged in substantial importation into North America by 1970, but their North American operations were of a volume corresponding to the "medium" size plant of the Commission study.

Given these estimates and descriptions, it is perhaps surprising that three of the top four largest tractor makers (Ford's tractors in the late sixties accounted for only about 5 percent of its total sales) have not been outstandingly profitable firms. Table 12-1 illustrates the profitability of most major North American farm machinery firms over the period relative to other U.S. corporations, and Table 12-2 shows the year-to-year profitability and asset earnings of the three

Table 12-1.   Average Return on Investment, 1946-67,[a] of Major Farm Machinery Firms Compared with That of 528 U.S. Manufacturing Companies for 1946-65[a] (Rate of Return on Equity—After Taxes)

ROI of Farm Machinery Firms[a] (1)		ROI Interval[a] (2) %		Approximate Percentage of 528 Firms in ROI Interval or Lower[a] (3) %
		20 and over		100
		18-20		87
		16-18		81
		14-16	central	73
Deere	11.0%	12-14	tendency:	60
Massey-Ferguson	10.3%	10-12	61% of	42
International Harvester	7.6%	8-10	firms[a]	25
Allis-Chalmers	6.3%	6-8		12
		4-6		5
J.I. Case	3.3%	2-4		2

[a]The two sets of data are for periods which do not exactly correspond. Scanlon's data were for the period 1946 to 1965. Since 1966 and 1967 (which were generally good years for business) were excluded from the figures for the 528 companies, it may cause the 1946-65 figures for the 528 firms to be slightly lower than the 1946-67 figures for the individual farm machinery companies. In that averages of roughly 20 years are being studied, however, the choice of terminal years should not materially affect the comparison.

Source: Martinusen and Barry, *Revenues, Costs, and Profits*, p. 36, based on Moody's *Industrial Manual*, various editions, and John J. Scanlon. "How Much Should a Corporation Earn?," *Harvard Business Review*, Jan.–Feb. 1967.

largest farm machinery firms by comparison with other U.S. firms. By comparison with either the profits of other firms or the long-term interest rate of probably about 5 to 6 percent, only Deere's profits appear to be notably "excessive." Unfortunately, profit figures with which to evaluate non-North American tractor-makers are not available because they are dominated by other activities in much the same way Ford's are.

Even if the RCFM manufacturing cost study is accepted as essentially sound, which this writer believes it to be, there are still a number of reasons why the manufacturing profitability estimates could differ so markedly from overall firm results: 1) The profitability calculations were done on the basis of only part of any company's tractor operations, manufacturing, on the basis of a rather arbitrary transfer price (despite the fact that it represented industry practice). The estimates of Table 3-6, however, show that over half of all assets involved in tractor manufacturing and wholesaling represent finished goods beyond the plant level. If one takes the estimates of Table 3-5 to be roughly correct, industry transfer price practice leaves non-plant operations with almost no margin of profit. The approximately 21 percent of dealer price represented

Table 12-2. Rates of Return: Net Profits After Taxes Divided by Net Worth (A) and Profits Before Interest After Taxes Divided by Total Assets (B)

	Deere A	Deere B	IH A	IH B	Massey A	Massey B	All Corporations A	All Corporations B^a	All Manufacturing Corporations A	All Manufacturing Corporations B^a
1946	7.4	5.1	5.3	4.0			9.9	5.8	10.3	8.5
1947	10.1	6.4	10.6	7.8			11.3	7.9	13.4	9.7
1948	15.9	9.4	11.8	8.3			11.4	7.9	13.4	9.7
1949	23.1	13.5	10.7	8.8			8.6	6.1	9.8	7.6
1950	22.1	13.6	10.9	9.0			11.3	7.6	13.4	9.6
1951	14.6	9.7	9.7	7.0			7.7	5.9	10.2	7.1
1952	12.3	8.6	8.2	5.5			7.2	5.0	8.0	5.8
1953	8.3	6.3	7.4	6.1			7.2	4.9	8.1	5.8
1954	6.7	5.3	5.1	4.3	6.0	4.5	7.1	4.5	7.4	5.4
1955	8.9	6.7	7.5	5.8	7.6	5.4	8.5	5.5	9.9	7.0
1956	6.1	5.0	6.5	5.3	1.9	2.7	7.9	5.2	8.6	6.3
1957	8.4	6.4	6.0	4.8	-2.7	0	7.1	4.7	7.7	5.7
1958	13.4	8.4	5.5	4.3	8.3	6.1	5.5	4.1	5.8	4.6
1959	13.9	8.9	9.3	7.0	10.7	6.2	6.5	4.8	7.8	5.8
1960	5.0	4.6	5.3	4.4	6.2	5.9	5.5	4.1	6.4	5.0
1961	8.6	6.8	4.7	4.0	6.6	5.9	5.7	4.2	6.3	4.0
1962	9.3	5.7	5.7	4.6	7.5	6.6	6.0	4.5	7.1	5.4
1963	10.9	7.8	6.8	5.1	9.4	7.5	6.1	4.9	7.8	5.7
1964	12.5	8.0	9.3	6.7	15.5	10.3	7.0	5.4	8.7	6.2
1965	10.0	7.2	9.0	6.5	12.4	8.5	8.4	6.0	10.3	7.1
1966	12.8	8.9	10.0	7.4	10.7	8.9	8.1	6.0	10.9	8.1
1967	8.9	6.9	8.2	6.4	6.2	6.6	7.3	5.4	9.0	6.4
1968	6.5	5.7	6.5	4.2	6.3	7.0	6.9	5.1	8.5	6.0
1969	8.0	6.7	5.5	4.9	7.1	7.2	5.6	4.7	6.9	5.3
1970	6.6	3.8	4.6	4.7	-5.0	3.0	4.4	4.3	4.9	4.5
Average	10.8	7.4	7.6	5.8	6.8	6.0	7.5	5.4	8.8	6.5

^{a}All non-financial corporations.

Source: Moody's *Industrial Manual,* various editions and U.S. Internal Revenue Service, *Statistics of Income*, various years.

by the difference between manufacturing transfer price ($4,019) and dealer price ($5,058) for an "average" tractor in 1968 would have just about covered estimated non-plant costs for the larger firms in the industry; 2) Tractors account for only about half the sales of a farm machinery firm's full line. Tractors and combines (another 15 percent by value) sell at higher margins than most other pieces of farm machinery [1]; 3) None of the major firms in the industry is solely a farm equipment producer. In 1967, 86 percent of Deere's sales were in farm machinery, and 78 percent of Massey-Ferguson's, but only 32 percent of International Harvester's (these percentages were all higher in previous years, however)[2]; 4) Even if figures for tractor profits for the largest firms were available, they would reflect to a substantial extent price-cost relations in countries other than the United States and Canada. The foreign-domestic breakdown of tractor sales by value is not available for individual companies, but in 1970, 17 percent of all of Deere's sales were outside North America, 19 percent of International Harvester's, and 66 percent of Massey-Ferguson's. Furthermore, the worldwide distribution of assets for these firms was comparable to sales.[3] The use of British components for half of the value of Massey-Ferguson's North American tractors should, of course, increase profitability, ceteris paribus; 5) The production facilities envisioned by the RCFM use the most up-to-date proven techniques and machinery, and, while it may well be argued that "the basic technology" has not changed since the early post-war period, the production equipment in the farm machinery industry was the oldest of all of the major subsectors investigated by *American Machinist* in 1968, with 41 percent of all machine tools over 20 years old and only 22 percent under 10 years old [4]; 6) Profitability rates were calculated for plants operating within 20 percent of designed capacity while, especially during the middle and late fifties, idle capacity may have diluted profits; 7) A final consideration is that book rates and internal rates of return are not the same thing. The slow growth of investment in new plant implied in the figures above resulted in large part from the low rate of growth of sales in the North American farm machinery industry. In constant dollars the 1950 sales level was not exceeded until 1958, and, although the last few years of the sixties saw a high level of sales, the annual compound rate of growth in constant dollars between 1950 and 1969 was only 2.2 percent by comparison with around 4.5 percent for manufacturing as a whole.[5] Measured book rates on largely depreciated capital, ceteris paribus, should give higher results than internal rate of return and would weigh in the opposite direction from factor 5.[6]

For all of the above reasons, there is very little point in a detailed examination of the aggregate profit experience of the industry leaders for comparison with RCFM results or as an approximation of the actual profitability of tractor operations. One is still left wondering if anything about the relative profitability figures of Table 12-1 is of interest in our evaluation of the tractor industry either in North America or elsewhere. The answer is that whatever information they convey is very indirect and must be interpreted with caution. The profitability figures in which farm machinery sales for the North American market loom largest are those of Deere (even here, however, only perhaps 40 to 50 percent of final sales are of tractors). It should be pointed out that although Deere's losses on foreign operations, mainly those on the Continent, were substantial in absolute amount, one estimate suggests that if the foreign ventures had broken even, they would have raised Deere's operating profit ratio only modestly over the period 1957 to 1967—from 14.8 to 15.5 percent.[7]

Massey-Ferguson's lower profitability, despite its low-cost source for tractors, the importance of which in its total sales is approximately the same as for Deere, must result largely from its rather distant third place in the North American market, its weakness in the corn belt, and the selling of most of its output at lower price-cost margins around the world.

International Harvester and Deere sell a comparable volume of farm machinery in North America, despite the fact that such sales are a far smaller percentage of total sales for the former firm. It is widely believed, however, that these sales are made at quite different levels of profitability; in other words, comparable farm machinery earnings are not being disguised by Harvester's relatively poor showing as a truck manufacturer. Industry opinion is that Deere's superior profitability can be attributed to three factors: 1) Deere's reputation for what *Forbes* has called "simple good management"[8] by contrast with International Harvester's almost opposite reputation noted in the previous chapter; 2) Deere's early entrenchment in the corn belt, which typically leads the country in the demand for large (and what are generally acceded to be high-margin) farm machines. For example, in the early seventies, data released by Deere indicated that while over three-quarters of its tractors sold in the United States were over 80 h.p., only 46 percent of International Harvester's were in this range; 3) low manufacturing costs [9] —which are in turn largely a result of conspicuously newer production equipment than most of its North American competitors and certainly than that of Harvester.

This writer was able to find only one piece of evidence bearing directly on the pay-off from tractor-making taken alone; it certainly confirms that high profits would result from the North American sale of tractors largely produced in a new, high-volume British plant. From an interview in 1968 with Bob Hampson, Ford vice-president and general manager of Tractor Operations, *Forbes* determined that "although Hampson won't divulge his own division's earnings he makes it clear he considers the performance of most farm machinery makers rather poor."[10] Elsewhere in the article it is stated, apparently in reflection of Hampson's analysis, that *"Tractors are more profitable than cars or trucks* [my italics] and big tractors are the most profitable of all."[11] One infers that tractor sales were very profitable because, although the previous year, 1967, had been a poor one for the firm, with an after-tax return of only 1.8 percent on equity, Ford's average after-tax profit rate for the decade 1957 to 1966 was 13 percent. During the year toward the end of which the Hampson interview was given, only about 40 percent of Ford tractors were sold in the U.S. and Canada, and Ford's machines were no larger than 67 h.p. Most of Ford's tractors were being sold in areas of lower profitability than North America and the firm was still very far from selling within North America a power-mix approaching that of the RCFM's "paper plant" study, although more powerful models were offered in the following year (see Figure 10-2).

MANUFACTURING EFFICIENCY

There are at least three questions to be asked about the manufacturing efficiency of the industry. First, how well has the industry achieved optimal plant size and utilization? Second, how well has it shifted production facilities to low-cost sources of supply? Third, how rapidly have changes in production technology been relative to the industry's potential?

Schwartzman estimated the cost savings available to the firms manufacturing in North America through consolidations of facilities and expansions of tractor output to 90,000 units. He attempted to estimate them by firm and found savings for Massey-Ferguson of 9 percent of unit costs, for Ford of 3 percent, and for International Harvester of 7 percent. From these savings must be subtracted 1 percent for additional transport costs. Deere's saving was estimated at 8 percent. For the smaller firms, Case and White, the estimates were much larger, 14 percent and 19 percent or more.[12] These figures are all lower than RCFM estimates for the shift from North American to British manufacture at any volume—this latter consideration is not discussed by Schwartzman at all.

Another approach to the estimation of possible manufacturing cost savings, assuming North American production and sales at the 90,000 unit level, is to find the transfer price into the distribution system at which a 10 percent internal rate of return on new manufacturing assets would be earned in the late sixties. The number found is $3,476 (by contrast with the $4,000 figure which actually prevailed). If one assumes that all post-manufacturing costs remain unaltered and that the same number and size distribution of machines would have been purchased if the savings per unit were simply passed on to the purchaser, the 1967 savings in the U.S. would have been $111,216,380, or 7.7 percent of the total wholesale value of tractors sold in that year. For Canada the figure would have been $15,662,536.[13] In fact, of course, unit sales would have increased, and, in addition, there would have been some savings on interest expense per unit in the distribution system.

As our previous discussion has illustrated, of the foreign makers, only Fiat ever realized an output substantially above that of the smallest tractor plant studied by the RCFM. Fiat was by 1970 producing in the range approaching that of the "medium" plant,[14] with some components, those used on crawlers similar in size to its wheeled output, produced at yet higher levels. For all other non-North American makers, one can assume that 20 percent would probably be a conservative estimate of the savings realized on most tractor components, although the large non-tractor engine output of British Leyland, Volvo, Renault, and the significant, but much smaller, truck engine production of Deutz somewhat complicates the picture for the approximately one-quarter of tractor costs which engines represent. Renault's lack of reliance on its own engines, however, and its outside purchase of engines from rather small domestic manufacturers,[15] indicates that the modification of automobile or truck engines necessary for use in tractors is sometimes substantial enough to eliminate the advantage which scale economies would otherwise afford.

Table 1–4, complemented by material subsequently developed about international movement of parts, suggests that in the mid-sixties only about 35 percent of world output came from plants judged to be of minimum efficient size, conservatively defined at 90,000 units per annum, and only half came from plants above about 60,000 units; all production above 60,000 was, of course, accounted for by the "big four." Further, 22 percent of 1966 output came from plants of less than 10,000 units, of which 14 percent originated in countries in the present study. Ignoring absolute cost considerations, the world industry must be given very low marks for production efficiency in terms of realizing economies of scale because the evidence points to the importance of such economies across a broad range of relative factor prices.

EXCESS CAPACITY

There is little available information about how the North American industry adjusted to the drop in demand in the early fifties, but if the industry demand projections noted in Chapter Two were taken seriously, and there is evidence to believe they were, then considerable excess capacity probably existed well into the sixties, despite efforts to get rid of redundant plant (International Harvester divested itself of over one-third of its total farm machinery plant between 1958 and 1962).[16] This excess capacity resulted not from some intrinsic structural characteristic, however, but rather from a serious misreading of long-run demand conditions during a specific period of time: the first decade after the war. There is no indication that serious excess capacity problems have been present outside of North America.

ADJUSTMENT TO INTERNATIONAL
DIFFERENCES IN PRODUCTION COSTS

In evaluating the performance of an industry, the price-cost margins, reflected in profit rates, of firms operating at minimum efficient scale provide a proxy for deviations from allocative efficiency (the equality of price and marginal cost), while the criterion of technical efficiency treats the extent to which the firms in an industry actually achieve minimum cost. The latter criterion has usually concentrated on the realization of scale economies, the avoidance of chronic excess capacity, and any evidence of operational "slack" in the setting of a closed economy. The discussion of Chapter One, however, suggested that another criterion of efficiency is also reasonable: the extent to which the production of an industry comes from areas of proven low absolute cost.[17]

This writer takes the position that the *prima facie* case for both national allocative efficiency and international specialization in the industry along the conventional lines of the theory of comparative advantage appears to be very strong and that the burden of proof lies with those suggesting that moving in such a direction is not warranted because of issues connected with what has come to be known as the "second best" problem.[18]

As documented in Chapter Three, there is a substantial and consistent body of evidence that costs were at least 25 percent lower in the U.K. than in the U.S. as far back as the early fifties. Further, this appears to be a very conservative estimate of the U.K. advantage following the devaluation of the pound in late 1967. A crude correction for relative cost increase in the early and mid-sixties might be made by interpolating linearly between a 25 percent cost advantage in 1960 and an 11 percent advantage on the eve of devaluation.

Table 12-3 shows rough estimates of the savings available to U.S. and Canadian buyers during various years if all tractors sold in North American had been produced at the estimated level of British costs and shipped across the Atlantic, with the savings passed on to the purchaser. Tractor production costs in North America are assumed to be one-half suggested retail price (SRP) and the "value of shipments" measure of the Bureau of the Census is assumed to be 65 percent of SRP,[19] so estimated savings of 25 percent of cost per unit can be estimated at 19.2 percent of "value of shipments" data. From this must be subtracted estimated transportation costs to the Midwest, where delivered prices were in fact quoted from machines of U.S. manufacture. These estimates, like the previous ones for the savings from more efficient U.S. production, simply assume that the same number and size distribution of machines would have been purchased, with manufacturing cost savings passed on to the North American buyer.

There are alternative estimation procedures for savings from U.K. production. One possibility is to use the prices of actual tractor models offered by Ford, Massey-Ferguson, and International Harvester in both the U.K. and the U.S., calculate non-production costs in the U.K., and estimate the savings available if the British tractor were introduced into the North American distribution system at a level which would put profits on a North American sale at the same level as those in Britain. For example, in 1961, the International Harvester B-275 had a suggested retail price of $2,912 delivered to Chicago. The British price (delivered to the U.K. purchaser) was $1,556. One can estimate that the company receipts in Britain were $1,276.[20] Because British distribution was direct and simple, the RCFM has estimated British selling, general, and administration costs at no more than two-thirds those in North America. Such expenses were approximately 11 percent of SRP in North America at the time, so British costs were about $213. To this must be added the cost of estimated British inventory carrying time, perhaps one month, or about $8. If these expenses are deducted from company receipts, the result is $1,055. A typical transfer price into the U.S. for such machines was about 61 percent of SRP or $1,779.[21] Declared freight costs were $168.[22] Accepting the entire margin between transfer price and net wholesale price in North America as unavoidable expense, $556 could still be taken off of both the North American transfer and net wholesale prices and the dollar amount left for profit would be brought down to the British level. (The estimated savings on this particular model are considerably higher than estimated for the same period using the procedure of Table 12-3.)

Table 12-3. Estimated North American Savings from British Manufacture, Various Years (U.S. Dollars)

	Plant Value Per Unit[a]	Percent Cost Savings	Amount Saved Per Unit	Transport Costs[c]	Net Savings Per Unit	U.S. Unit Sales	U.S. Total Savings	Canadian Unit Sales[d]	Canadian Total Savings
1955	$1,589	19.2%	$305	$197	$108	268,100	$ 28,954,800	25,825	$ 2,789,100
1960	$2,291	19.2%[b]	$440	$240	$200	177,949	$ 35,589,800	25,399	$ 5,079,800
1965	$3,398	11.5%[b]	$391	$193	$198	197,720	$ 39,148,560	26,855	$ 5,317,290
1968	$4,430	19.2%	$850	$164	$686	195,435	$134,068,410	23,098	$15,845,228

[a]Unit values calculated from U.S. Census data reproduced in *Automotive Industries*, various issues.

[b]By 1965 production cost advantage estimated to be only 15 percent.

[c]Shipping costs estimated from IH's declared shipping costs in 1961. NRFEA, *Official Guide*, 1961, p. 98, and *Final Report*, p. 276.

[d]Farm only.

One could proceed from a couple of models for each year, or for selected years, and, assuming that other major firms would have to match the Ford, Massey-Ferguson, and International Harvester prices or withdraw from the section of the market in question, estimate the savings for *all* tractors sold in the horsepower classes for which the matching was done. Although increased precision would appear to be achieved, the gain would be largely specious. Estimated gains would vary by model and company, and the exact figure to use for transfer into the North American distribution system is to a large extent arbitrary. Further, as the exercise is illustrative only, there is no point in delimiting savings estimates to the horsepower categories in which there actually was British production of whole tractors. In the event, the models for which comparisons are available cover between 45 and 55 percent of the total value of tractors sold during any year in North America up to 1970. Nevertheless, nowhere in its extensive research did the RCFM find any evidence that North America has even a comparative, let alone an absolute, advantage in the production of larger machines. It is therefore of interest to consider how much saving might have been realized by assuming all machines to be made in the U.K. and adjusting transfer prices of tractors sold in North America so as to yield only a 10 percent internal rate of return on new manufacturing assets of efficient scale. As determined in the preceding discussion, the transfer price yielding a 10 percent internal rate of return on new 90,000-unit plant in North America, treating the entire margin between transfer price and net wholesale price as unavoidable cost, was $3,476. If, by assumption, all British costs are treated simply as those of North America proportionately "scaled down," the same internal rate of return on British assets would be achieved with a transfer price bearing the same percentage relation to estimated British manufacturing costs as the North American price yielding 10 percent does to estimated North American costs. If the seemingly conservative estimate of 75 percent of U.S. production cost is used for the immediate post-1967 devaluation period, the British cost estimate is $2,412, and the receipts necessary for a 10 percent internal rate of return on manufacturing assets would be generated by a transfer price of $2,606. Actual transfer price in North America was $4,000, so, allowing $164 for additional transport costs (see Table 12-3), the potential reduction in transfer and wholesale price per tractor is $4,000−($2,606 + $164)=$1,230 savings per machine. This figure was 24 percent of the wholesale price of an average tractor in North America in 1968 and represents a potential saving of $240,385,000 for the U.S. and $31,128,840 for Canada, on the number of units actually sold in 1968.

As was implied in the previous chapter, one would never expect to find international migration of even a strictly profit-maximizing firm if capital costs were truly "sunk" and average variable cost in the original location were less than average total cost in another area. Chapter Three pointed out, however, that in tractor manufacturing, over 70 percent of all costs are for material, and capital costs are never more than 15 percent. Manpower expense of one kind or another accounts for the remainder. There is no reason to believe that the relation of capital to total costs in the industry has changed substantially over time. It thus appears reasonable to conclude that, even if capital costs were truly "sunk," it would have been profitable for the North American manufacturers to have moved their operations abroad at least by the late fifties. This would, of course, have entailed expansion of the operations of foreign supplier fiirms and a contraction of supplying firms in the United States. There is no indication that the purchase of parts and material obtained from suppliers could not at any point have been accomplished at the same cost to the migrating firm as for those firms already buying (the RCFM's rough estimate was that actual outside purchased parts in Britain were about 25 percent cheaper than their U.S. counterparts after the devaluation of 1967).[23] It also appears likely that U.S. suppliers had little capital sunk in serving the tractor industry. For most suppliers the latter was probably a rather minor market for their output, and the demand for self-propelled machinery was steadily expanding. Furthermore, it seems very conservative to assume that all capital costs in tractor-making itself were, or are, irredeemably sunk. Buildings are saleable (but account for a minor part of capital expense), and most production equipment can be moved. Massey-Ferguson, for example, has considered the possibility of shifting a substantial amount of its production equipment within North America.[24]

If the United Kingdom and Italy are treated as the only two countries in the study showing sustained evidence of being low "absolute cost" areas (and Italy is held to be so on the basis of internal prices rather than direct production cost estimates), then by yet another criterion, the world industry appeared to be performing poorly in 1970. Although production from these countries loomed large in the markets of non-producing countries, United Kingdom and Italian plants provided only 32 percent of estimated non-Communist production in the mid-sixties and about the same fraction in 1970.

While the failure of the industry to adjust its production internationally by 1970 is sufficient for an indictment of its performance, it is still of interest to consider whether or not the U.K. and Italy are likely to retain their positions of absolute cost advantage. The early seventies were a period of great inflation and uncertainty about exchange rates, and one cannot predict with absolute confidence that, when a more stable situation develops, the U.K. and Italy will have retained their advantage. Conditions contributing to the ongoing advantage of these countries include the absence of a permanent and substantial increase in tractor prices (assuming they reflect costs) relative to the domestic price level by comparison with other countries, and the necessity for exchange rate equilibrium of an overall price level relative to other countries not significantly higher than it was prior to the disturbed period when expressed in international prices. A factor which could affect the former condition might be the increased bargaining strength of labor. Factors which could cause the second condition not to hold include: disequilibrium in the base period, tariff changes, changes in capital flows, and differential rates of real economic growth.[25] Taking all of these considerations into account, there was no prospect as late as the end of 1973 that the U.K. and Italy were losing their position as locations of lowest absolute cost by a substantial margin among the countries in the present study.

PROGRESSIVENESS IN MANUFACTURE

An estimate of productivity change in terms of output per man hour for the U.S. farm machinery industry as a whole was made early in 1972 in connection with the work of the Price Commission, and it is the only one known to the writer for any country. The annual rate of productivity increase averaged 2.4 percent over the 12 years from 1958 to 1969. This can be compared with the weighted-by-sales manufacturing average of 3.6 percent and the figure for "Motor Vehicles" of 4.1 percent.[26] Why has the rate of productivity advance been so slow relative to manufacturing in general and to a somewhat similar industry? An outstanding characteristic of farm machinery which probably provides much of the explanation is the increasing age of machine tools used in production. There has been only modest plant expansion of any kind in the North American industry, and no wholly new tractor manufacturing facilities have been built since 1950.[27] Furthermore, despite the fact that the decline in the industry's output after the mid-fifties presumably resulted in divestiture of the most antiquated plant and machinery, the increasing age of equipment in use becomes really marked from the late fifties onward.

Table 12-4 shows the amount of new metalcutting equipment in the farm machinery industry by comparison with other industries for 10 periods from 1925 through 1968. Metalcutting equipment accounted for over 70 percent of the total number of pieces (no other relative measures are available) in farm machinery manufacture, and it is safe to assume that it looms considerably larger in tractor-making than in the production of most other farm machines. The percentage of new metalcutting equipment was smaller in 1968 than at any previous time since shortly after the tractor was first mass-produced.

One can only speculate about an explanation for the relative absence of new production machinery. Excess capacity may have reduced wear on some equipment and lengthened its physical life for all of the participating firms. Factors which might have contributed to increasing the economic life of the equipment could have been an unusually low elasticity of substitution of capital for labor in the industry and (or in addition to) a slow rate of advance in the industries supplying the production equipment for farm machinery. Given the similarity of the production processes for at least half the farm machinery industry's output (i.e., self-propelled machinery) to those in the far more rapidly progressing motor vehicle industry, one may doubt the importance of the latter two factors, despite the vast differences in volume between the industries. Furthermore, there is abundant evidence of the increased performance of the latest designs of

Table 12–4. Metalcutting Machine Tools: Age of Machinery in Sixteen Industries, Selected Years

	Percentages Under 10 Years Old									
	1925	*1930*	*1935*	*1940*	*1945*	*1949*	*1953*	*1958*	*1963*	*1968*
Farm machinery	35	42	40	31	45	50	50	28	27	22
Construction, mining, materials handling	—	51	26	27	51	52	42	36[1]	26	30
Metalworking machinery	—	—	34	29	55	57	40	38	37	40
Special-industry machinery	44	55	24	26	43	46	38	29	28	32
General industrial equipment	48	53	25	22	46	56	45	34[2]	31	33
Fabricated metal products	41	51	30	24	53	55	48	41	36	37
Office and business machines	—	47	27	31	42	52	47	34	43	40
Service industry machinery	—	—	—	—	—	—	—	—	31	30
Electrical equipment	55	49	35	30	53	63	49	43	41	39
Communications equipment	—	54	40	31	53	62	57	58	61	52
Household appliances	—	—	45	37	63	44	43	34	20	26
Motor vehicles & parts	73	73	—	—	—	—	—	54	32	33
Complete aircraft	—	—	—	71	98	84	55	50	33	37
Aircraft engines & parts	—	97	87	39	64	65	49	50	44	31
Precision mechanisms	—	45	—	23	60	62	55	41	44	41
Ordnance, shipbuilding, railroads	56	48	33	27	62	57	36	28	30	28
Metalcutting average	—	—	—	—	—	—	45	40	36	37

Notes:

Percentages in some inventories have been adjusted, so that data will be comparable. Over the years, changes in SIC industry codes make it impossible to classify some industries exactly, especially in the earlier inventories. Blanks mean no comparable data are available.

1. Figures before 1958 do not include Materials Handling Equipment Industry.

2. 1958 figure includes Engines and Turbines industry, which is not included with General Industrial Equipment in the other inventories.

Source: *American Machinist.* "The Tenth Annual American Machinist Inventory of Metalworking Equipment: 1968."

many of the pieces of equipment used in the production of tractors over the years.[28]

A more important explanation for the increasing age of production equipment probably relates to the inability of firms to raise capital for modernization or their unwillingness to commit new plant for North American manufacture. Over 20 percent of tractor output and a similar fraction of the production of other major farm machines over the period were in the hands of companies which were barely staying above water and which certainly did not have enough strength in capital markets to engage in substantial renovation of facilities. A related consideration, which would have retarded both the willingness of the small firms to invest and perhaps some potential sources of finance to provide capital, might have been related to concern about the direct or indirect impact of foreign competition on the prospects of the firms. Deere's fixed capital expenditure as a percentage of sales averaged 5.3 percent over the period of 1957 to 1967, with a considerable (but unknown) amount going to its integrated tractor plant at Waterloo. By contrast, the Case average (55 percent of the 1967 sales of which were in farm machinery) was only 3.3 percent. Allis-Chalmers (27 percent farm machinery sales in 1967) averaged about 2.0 percent over the period, and White Motor and predecessor companies (about 25 percent of the sales of which were farm machinery after the acquisitions) averaged well under 2.0 percent. (Massey-Ferguson, which had over half its assets outside of North America, had a rate of increase of 5.5 percent).[29]

Far more surprising than the low rate of investment of the smaller companies, however, was the low outlay rate of International Harvester: it averaged only 3.9 percent over the 11-year period.[30] Furthermore, there is qualitative evidence that this low average does not disguise a usually high rate of investment in farm machinery production. The truth may be the opposite. In the late sixties, for example, Harvester was producing tractors from four separate assembly plants which themselves relied on components from very old factories.[31] Why? The sunk cost answer is as unpersuasive for purposes of explaining low domestic investment as it was for explaining the lack of international migration of production. Deere found it profitable to engage in piecemeal but continuous modernization of its tractor facilities, despite a legacy of antiquated excess capacity; why should Harvester not have done likewise? One cannot say with certainty, but a plausible reason for Harvester's reluctance to invest may have been doubts about the most appropriate geographical distribution of plant for tractor (and perhaps other farm machinery) manufacture. President Harry Bercher predicted in 1962 that an increasing volume of tractor production for the North American market would be coming from abroad and that ". . . this international operational flexibility in itself should give Harvester a considerable advantage over its more parochial U.S. competitors. Deere, for example. . . ."[32] While the company may not have devised a master plan for international production by the end of the period under review, the desire to preserve options may have contributed to a postponement of a major retooling of tractor facilities at home.[33]

Using the rate of increase in labor productivity as a measure of progressiveness in manufacturing, the North American industry appears to have done poorly, and this in turn seems to be largely a function of the industry's failure to employ modern production equipment. Unfortunately, no comparable information is available for other national industries, but much of British and Italian production equipment is far newer than that in North America. Ford, David Brown, and British Leyland all constructed substantial new British plant during the sixties, and in the Ford case, the renovation of plant was close to complete for nearly all of the production machinery necessary for the maximum in-house production discussed in Chapter Three. Deutz built a completely new plant in 1961, and Fiat's rapidly expanding output in the sixties came largely from new plant.[34]

PRODUCT DEVELOPMENT AND QUALITY

Most discussions of product development and quality lack suitable benchmarks and can easily lapse into a speculation on what might have been. An important attribute of a multicountry study under some circumstances might be the provision of realistic standards of comparison, both statically and over time. Unfortunately, this approach does not prove to be very fruitful for our purposes because the same actors were producing similar machines after at least the late fifties for sale in all of the countries in the study, and the extent to which differing machines produced earlier adequately reflected differing demand conditions is not well understood. We were able to conclude with confidence that the speed with which the Ferguson system was diffused in North America was unimpressive.

It has been stressed in this study that really major technical improvements in tractor design since the war have been nil if one properly views the Ferguson system as only being diffused and improved upon during this period.[35] On the other hand, counterfactual speculation about the possibilities of tractor design is difficult. It appears that the improvement in farm machinery before World War II, so much admired by Schumpeter,[36] has continued apace with some types of equipment while tractor design virtually stagnated. Once the Ferguson system was completely diffused, however, informed opinion has not been able to point to important areas of design neglect (leaving aside comfort and safety for the moment); the lack of advance is usually assigned to the rather simple nature of the product and the limited number of areas of possible design improvement. Continuing refinement of the hydraulic system has probably been the single most important area of advance. Otherwise, the products of the world industry appear to have improved in various minor ways while maintaining or increasing their durability. An example of improvement has been the establishment of substantial commonality of parts within the lines offered by most manufacturers,[37] thus in part countering the increased manufacturing cost and repair parts provision problems threatened by a larger number of models. Fuel economy also increased slightly over the years, at least in North America.[38]

If one focuses attention on production quality control or the durability of the product, international comparisons seem to afford increased scope for the evaluation of firms and national industries. However, with the exception of some quality problems mentioned in Chapter Nine in connection with particular models, there is no systematic information about differences among the machines of different makers in the various markets gathered over significant periods of time on either quality control or durability. On the other hand, a strong *prima facie* case that international convergence by both criteria had been accomplished by the late fifties or early sixties is provided by the practice, detailed earlier, of the small U.S. full-line firms importing David Brown, Fiat, and Renault machines to be sold under their own names with no mechanical modification.

For production quality control, the original Ferguson machine produced in Britain by Standard Motors would appear to have been difficult to match. In 1957, after an output of over a half million machines, warranty claims were less than 15¢ per tractor.[39] This, of course, was the product of only one manufacturer, but an indication that the *general* level of quality may have continued to improve, at least in Britain, is a 1967 NFU statement that "the number of farmer complaints about faulty machines . . . has diminished in recent years." [40]

If declared warranty policies are any guide, it appears that, whether or not production reliability improved, at least the manufacturer's willingness to back up his machine generally did. A full-year warranty has been standard throughout the period for machines from most manufacturers selling in North America, the United Kingdom, and Australia. Machines sold on the Continent in the late fifties typically had six or nine month warranties;[41] by the mid-sixties these had been brought up to what had become virtually a worldwide one-year standard.

MODEL PROLIFERATION AND STYLING CHANGE

The discussion of Chapter Ten was skeptical of the argument that model proliferation in North America has been excessive, although correcting for the range of horsepower offered did indicate international differences in the "density" of offerings. The welfare implications of such differences are not clear, however. It may be that the increased offerings met real customer needs to a greater extent than they increased cost; the opposite may also be the case.[42] In the event, the relatively low levels of concentration with which the proliferation of models was associated implied operation of highly suboptimal plant by nearly all market participants; this, however, is an analytically distinct issue.

Although it was determined that the industry is largely free of costly and spurious product change of a kind clearly geared to stimulate sales and divorced from other improvements, it was also discovered that a considerable amount of the interfirm variation in parts specification does contribute to increased repair parts expense with no compensating benefit to the purchaser, and it appears that the industry has considerable incentive to maintain the status quo.

COMFORT AND SAFETY

Where comfort and safety have gone hand in hand in such a fashion as to be readily recognizable by the purchaser, the industry has often responded. The development of the tractor seat from the old metal scoop-on-a-spring to a more comfortable design resulted in a diminution in fatigue and spinal disorders and presumably in an increase in safety of operation as well.[43] Both comfort and safety are also provided by most tractor cabs. Many cabs are especially designed to provide rollover protection, but the 20 percent of all tractors sold in the United States in 1970 with cabs appear to have been purchased mainly because they afforded protection from dust and allowed for heating and air conditioning.[44] In North America, and apparently in most other countries as well, farmer sentiment about the need for health and safety features per se has never been intense, and the generally conservative farm community has failed to produce even the vocal minority characteristic of the users of other products. Consequently, there appear to be many areas in which the machinery makers have failed to take even the most minimal initiative. Simple and inexpensive steps to reduce tractor noise, for example, were long delayed despite evidence that permanent hearing damage was being caused. The companies are believed to have thought that the farmer associated noise with power.[45] Industry interest increased considerably after the Nebraska Tests began to include noise emission in 1970; the OECD began similar testing in the same year. An even simpler example of industry neglect is the heavy springing of the tractor clutch which makes it sometimes difficult to depress in emergencies. The companies are alleged to believe that heavy springing suggests quality of construction to the farmer.[46] Information is unavailable about the extent to which industry inattention to cheap and minor safety features prevails outside of North America; the only evidence in indirect: critics of North American practice have never pointed to superior voluntary performance by sellers in other countries.

A great difficulty in discussing tractor safety is that there is no accurate reporting of non-fatal accidents, and the most pertinent data surrounding fatalities are not always recorded with care. Nonetheless, it has long been known that the principal cause of work-related rural fatalities is the overturning of tractors which accounts for as many as 600 deaths in the U.S. per year. This was a major factor which led to a much-publicized (and statistically controvertible) 1969 Nader Report conclusion that farming was the "third most dangerous occupation" after mining and construction.[47] It has been established from Swedish data that nearly all such fatalities can be eliminated by some form of rollover protection which Sweden first made mandatory in 1959. Such protection, however, adds between $200 and $1,500 to the price of the tractor, depending on whether the method employed is a rollbar or a safety cab.[48] While tractor overturns have been officially recognized as a major source of rural danger in every country in the study, conpulsory regulation does not appear to have been generally popular with farmers themselves. The British National Farmers Union fought compulsory cab legislation all during the sixties—ultimately to no avail; Britain introduced legislation making safety frames mandatory on new machines from September 1, 1970, and during that year similar standards were being studied on the Continent by the OECD.[49]

What may be an example of North American industry action to avoid government regulation was the adoption by the American Society of Agricultural Engineers (ASAE) in 1969 of a crash program to reduce accidental rural deaths during the 1970's by 50 percent.[50] The program was set up to study all aspects of farm safety outside of the home and to convey the results of its findings to all interested parties. Machinery deaths, most of them from tractor accidents, account for over 40 percent of the fatalities, and the safety drive was suggested to the ASAE by a Ford executive while legislation was being prepared in the Congress to authorize an investigation of tractor safety by the Department of Transportation. During the course of the investigation, the industry pointed with pride to its ambitious efforts to cut down fatalities and injuries.[51] Although the industry's sincerity about improving rural safety is not in question, some form of compulsory rollover protection was thought to be one of the most likely outcomes of the Department of Transportation's investigation, and this was probably regarded as a profit-reducing requirement by the industry because protection of this kind was demanded by only a small minority of users.

When the manufacturers were questioned about their own views concerning mandatory safety equipment, the investigators concluded that, "Their reasoning, as near as could be determined, is that the farmer is certain to oppose such rule-making action, and manufacturers would not want to be on record as endorsing such specification."[52] While the investigators were naive in assuming that this was the industry's prime motivation because "it stood to make a profit on the requirement,"[53] there can be little doubt that it adequately gauged rural opinion. In a national survey conducted by a farm magazine (and verified by testimony given before the RCFM at about the same time), it was discovered that less than half of the farm community thought its equipment less safe than it should be, and most dissidents held that a rollbar was the principal feature lacking.[54] Rollbars were available as optional equipment, and those farmers interested would presumably buy one on their next tractor. There was no sentiment for government regulation.

The industry's stand is scarcely surprising; government regulation was not popular with those most directly affected, and the industry might well have believed that the government's probable maximum proposal would be virtually the same as the industry's minimum substantive accommodation: non-optional rollbars. Cabs were much more expensive, and their additional contribution to safety not firmly established. This suggests a general proposition: The popularity of proposed government regulation among those directly affected may help determine the industry's strategy toward the regulatory agency; it will expect milder regulatory response when the provisions are not generally popular, particularly in the absence of intense minority lobbying. This is nothing more than the assumption that the regulatory agency is unlikely to seek trouble when there are no compensating pay-offs.

When questioned about rollover protection during the Transportation safety hearings, the manufacturers pointed out that, although rollbar and cab standards had been developed under industry auspices in the late sixties, and that all firms offered both rollbars and cabs as optional accessories, the industry believed it would be in violation of the antitrust laws if it moved jointly to make such protection standard equipment. The Transportation report continues: "Whether real or imagined, the fear appears to be that such agreement could be interpreted as price-fixing, or as constituting a move to force smaller, less well-prepared manufacturers out of business."[55] It may well be true that this industry is particularly sensitive about antitrust issues, but in light of virtually simultaneous moves in other industries to adopt non-optional safety equipment and the fact that all of the tractor-makers were already offering the equipment optionally (as the report itself acknowledges elsewhere), antitrust grounds for inaction are hard to take seriously.

If there is an antitrust issue involved in making such safety equipment non-optional, it would appear to relate to the prohibition of tying in Section 3 of the Clayton Act. Any piece of optional equipment available not solely from producers within an industry may become popular enough that the industry's ability to appropriate the entire market outweighs what it would lose by not being able to offer some customers the product without the special equipment. Injury to independent manufacturers of the equipment might then provide a legal ground for complaint.

When safety equipment is involved, however, the manufacturers might seek an additional justification for their actions, and public or private support of mandatory factory installation could be their vehicle. Such a legal requirement might look attractive to regulatory agencies, because it would obviate the expensive (for the regulators) and less effective methods of either monitoring at the point of sale or inspecting equipment in use. It must be stressed that the hypothesis above relates to the *type* of safety equipment being considered in the farm machinery industry, but only in part to the structural conditions there. It does not explain any actions yet observed in the industry.

The industry's behavior in the area of safety, at least in North America, has scarcely been innovative, and some inaction such as that related to noise and clutch springing appears quite damning; nontheless, the failure to adopt expensive non-optional safety devices not generally favored by customers is understandable.

DISTRIBUTION

Chapter Three presented estimates of the savings which would be gained under prevailing North American distribution practices by the provision of equipment at the scale of the largest firms. It was determined that approximately the same economies of scale, around 20 percent, would attach to distribution expansion from 20,000 to 90,000 units as would accrue to a production expansion. None of the material gathered for this study indicated that the degree of farmer satisfaction with the distribution system for tractors and other farm machinery differed widely from country to country. This fact suggests the fundamental question: Could the North American (and perhaps the Australian) farmer be equally well (or better) served by a cheaper distribution system?

Selling and General Expenses

One place to begin looking for possible sources of saving is explicit "selling, general, and other expenses" as a percentage of sales revenues for farm machinery as compared with those for somewhat similar manufacturing industries. At first glance, such a comparison seems significant. Over the period of 1956 to 1969 General Motors spent, on average, only 4.9 percent of sales revenue in this way; the similar figure for Deere was 11.6 percent. When these figures are corrected for sales volume, however, they are inconclusive (see Figure 12-1). While it may be that there is a tendency for the farm machinery industry to incur higher expenses in this category, the relation between size of self-propelled machinery producer and expense percentage appears to be more interesting. (Direct advertising for all firms in both industries was less than 2 percent of sales.) Most of the sales range for automobiles is, of course, far beyond that which could be achieved by firms in the tractor or farm machinery industry without vastly higher levels of world concentration than presently obtain. Something close to a world monopoly in farm machinery would have been necessary to attain even the size of the Chrysler Corporation.[56]

A striking contrast in this cost category is provided by the experience of Versatile and CCIL. Both tiny Canadian firms (Versatile $15.6 million and CCIL $18.6 million (U.S.) in 1966) relied on a simple distribution system, and both had astonishingly lower costs than their competitors. Between 1963 and 1967 the Versatile figure was 3.8 percent and that for CCIL 2.7 percent. Versatile's

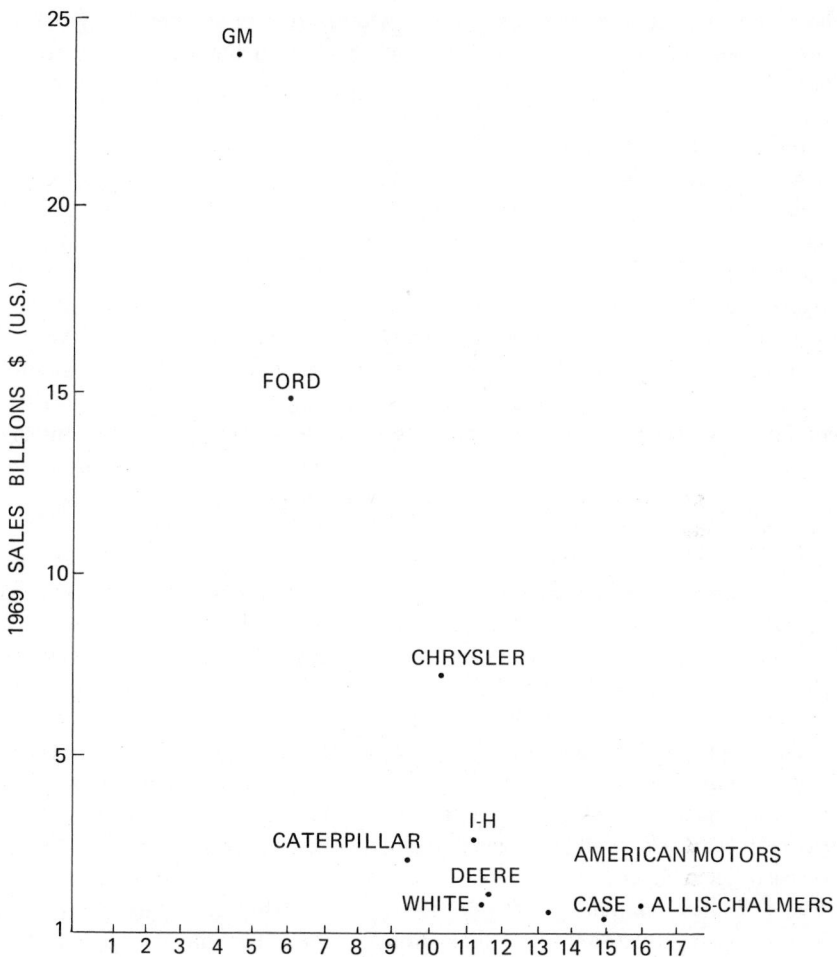

Selling, General, and Administrative Expenses as a Percentage of Total Sales—
1969

Figure 12-1. The Relation Between Selling, General, and Administrative Expenses and Sales—Selected Manufacturing Firms—1969[a]

[a]1969 is the last year for which this information is shown separately for J.I. Case rather than for the parent Tenneco Corporation.
Source: Moody's Industrial Manual.

subsequent conversion to floor-planning, noted in Chapter Seven, undoubtedly generated a growth in this kind of expense because company supervision would at the very least be necessary to supervise account payments, and, after the time of the latest available data, CCIL also reorganized to give more attention to its 60 large depots through an equivalent of blockmen.[57] Nevertheless, whatever developments followed, there was a very wide margin to be eaten away before these unusual distribution systems approached those of the major companies in cost.

Inventories and Receivables

A much more interesting picture of the major firms in the industry is obtained when one considers the relationship of receivables and inventories to final sales. Figure 12-2 illustrates the ratio of receivables to final sales and the ratio of receivables plus inventories to sales over the period of 1950 to 1970 for the two largest predominantly farm machinery firms in North America and the two smallest U.S. automakers. The farm machinery companies had much higher ratios in both categories, but the greater difference was in the receivables ratio. It was the practice for all four companies to assist the final purchase of their equipment through non-consolidated subsidiaries which are not included in the figures shown, but in contrast with automobile practice, dealer credit is an important use of capital in farm machinery, and for Deere and Massey-Ferguson this credit nearly exhausts the "receivables" entry in the main company books.

Year-end receivables plus inventory was 13 percent of Chrysler's 1970 sales and 28 percent of American's, while the Deere percentage was 97—at the closing of the company's books nearly a year's sales were "in the pipeline" somewhere between the end of the assembly line and the customer. For Massey-Ferguson, which had 66 percent of its sales outside of North America in contrast with Deere's 17 percent, the receivables plus inventory figure was 72 percent.

The exact role of tractors in the very substantial volume of machinery between manufacture and final sales within the broader industry is not known, but there is certainly little doubt that it is extremely important. Table 12-5 gives tractor inventories at all levels including dealerships on an average annual basis for each calendar year from 1955 (when figures were first estimated) through 1968 and their relation to unit sales in that year. Although average figures for the sixties are noticeably lower than those for the fifties, the average monthly inventory remained very large, and there is no definite downward trend after 1962. (The Bureau of the Census Series upon which the figures after 1959 are based, and which receives financial support from the industry, stopped presenting inventory figures after 1968.)

Chapters Two and Three stressed the differences in production practices and seasonality of sales between most farm machinery and tractors. Tractors differ little in seasonal variation of sales from automobiles (although with a different pattern) and are produced year-round in separate plants, while much other machinery is highly seasonal and shares common plant. While it is therefore not easy to interpret the significance of large inventories and receivables for farm machinery in general, the situation for tractors is otherwise. In tractor production and distribution there is little reason to believe that production problems rather than promotional activity should produce a volume of inventories and re-

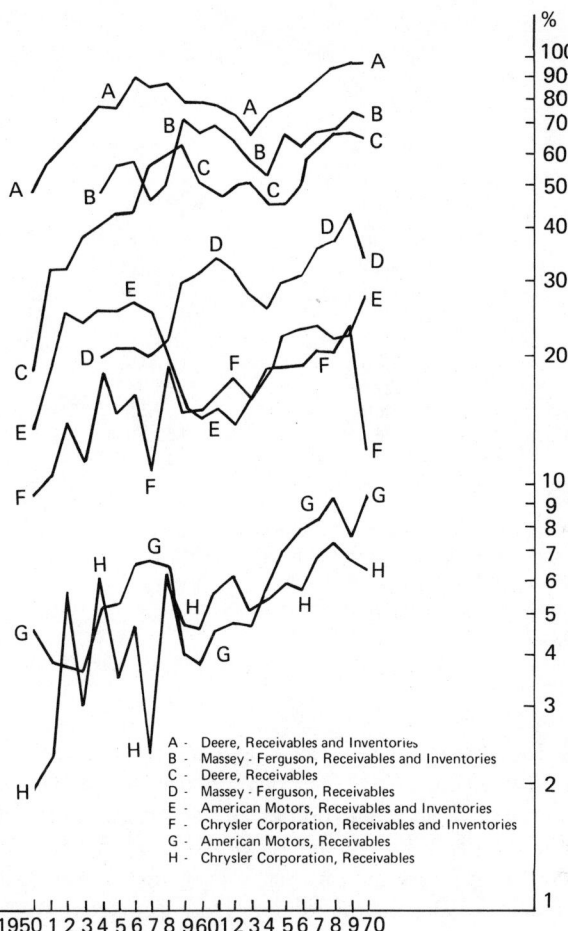

%
100
90
80
70
60
50
40
30
20
10
9
8
7
6
5
4
3
2

A - Deere, Receivables and Inventories
B - Massey - Ferguson, Receivables and Inventories
C - Deere, Receivables
D - Massey - Ferguson, Receivables
E - American Motors, Receivables and Inventories
F - Chrysler Corporation, Receivables and Inventories
G - American Motors, Receivables
H - Chrysler Corporation, Receivables

1950 1 2 3 4 5 6 7 8 9 60 1 2 3 4 5 6 7 8 9 70

Figure 12-2. Receivables and Receivables Plus Inventories As a Percentage of Annual Sales-Selected Firms 1950–1970
Source: Moody's *Industrial Manual,* various editions

ceivables significantly different from that prevailing in the automobile industry. Nevertheless, when one compares the year-end inventory plus receivables for the smallest U.S. auto-maker (which is not unrepresentative of the figure over most of the year) with comparable figures for tractor distribution, the results are very different. For example, in 1968, the most recent available year for tractors, American Motors' total inventories plus receivables was 22 percent; the average that year for tractors was just over 68 percent. Using the approximate wholesale unit value of $5,000 (based on actual industry experience), cutting outstanding tractor inventories by two-thirds would have released about a half billion dollars

Table 12-5. Inventories and Sales

	Sales	Average Monthly Inventory	Ratio
1955	268,100	167,400	62.4
1956	207,700	189,000	91.0
1957	199,000	162,600	81.7
1958	216,000	155,500	72.0
1959	208,600	160,500	76.9
1960	177,949	153,100	86.0
1961	166,770	129,740	77.8
1962	183,244	114,951	62.7
1963	186,148	114,144	61.3
1964	192,076	107,493	56.0
1965	197,720	96,967[a]	49.0[a]
1966	227,000	119,582	52.7
1967	212,245	135,501	63.8
1968	195,435	133,129	68.1

Notes:
1. Before 1960 estimates of inventories are from *Implement and Tractor* and are for whole goods at all stages of distribution. Census figures after 1960 are for all units made in the U.S. Beginning in 1965 finished goods at factory excluded.
2. After 1965, estimates are for end of month rather than first of month.
[a]First 8 months only.

Source: *Implement and Tractor*, various issues. Bureau of the Census, *Current Industrial Reports*, Series M35S, various years.

of capital in North America in 1968; cutting them by 50 percent would have released a third of a billion. The former figure is over 65 percent of the capital necessary for optimum scale production of all tractors sold in North America in 1968, based on estimates from Chapter Three. Half to three-quarters of a year's sales perpetually in inventory appears highly excessive in an industry in which annual unit demand seldom fluctuates more than 20 percent around the mean and for which, as established in Chapter Three, cost changes associated with a 20 percent departure in output volume from a plant of a certain designed capacity are modest and which are reductions for a production increase.

Dealership Size and Service

It has been demonstrated that the high costs of North American distribution are determined in large part by the number of dealerships. In Chapter Five it was argued that dealership proliferation was an early form of competition, but one from which the industry, led by Deere, was retreating. Nevertheless, it was established that to be efficient as purveyors of wholegoods, retail outlets would have to be much larger on average than they were in the late sixties. On this count at least, it is clear that there are too many dealerships. Our entire discussion, however, has stressed the importance of after-sales service in the product "package" delivered to the farmer. What would be the impact on after-sales service of dealership consolidation?

Extremely close attention to this issue by the RCFM found virtually all of the evidence on the side of the superior performance of larger dealers, and the *Report* comes down strongly for fewer, larger dealers on the basis of service alone.[58] Farmer complaints about service were strongly skewed toward the smaller dealerships, while larger dealerships typically complement their more sophisticated in-house equipment with highly mobile field facilities (sometimes including airplanes and helicopters) and large low-bed trailers to move machinery when necessary. The pages of *Implement and Tractor* over the past decade have been rife with the experiences of dealers successfully attempting to marry economies in wholegoods selling and service facilities with the ability to deliver fast, dependable service. In some instances, the establishment of a permanent parts and service facility (such as the John Deere "service center"), closer to the farmer than the wholegoods outlet, may be warranted. The only really difficult problems of service arise in thinly farmed areas where dealerships might indeed have to be inconveniently located to be of economic size and where service would be best provided by "service stations," perhaps authorized by a number of companies to sell parts and perform warranty work.

Because of the increased profitability which usually accompanies larger dealerships, well-established dealers will often combine their operations with the cooperation of the manufacturer. On the other hand, where such an amalgamation is not feasible because of reluctance on the part of one or more of the parties, the manufacturer is understandably careful about forcing the issue and perhaps driving one of the dealers (assuming he is running a creditable if suboptimal operation) into the arms of another supplier.

Despite the fact that in many localities, large-scale wholegoods selling and after-sales service have proven successful and the considerable incentive manufacturers have to consolidate their dealerships due to economies in wholegoods handling and the increased sophistication of service available from large dealers, the competitive importance of the display function will probably continue to keep alive inefficiently small dealerships for decades if present trends continue. The following chapter suggests one line of policy which might induce the manufacturers to move more rapidly in the desired direction.

Distribution Expenses in Detail

A detailed look at distribution expenses in the North American farm machinery industry is provided by Canadian data gathered by the RCFM (see Table 12-6). Post-production expenses of major firms in the industry were disaggregated, and, in addition, the implicit cost of inventory held either at branch warehouses or on dealers' premises was estimated. Central administration costs were judged to be between 2 and 3 percent of total sales, and they are not included in

Table 12-6. Branch Office Marketing Costs as Percentage of Net Sales, Nine Major Full-Line and Long-Line Farm Machinery Manufacturers, Canada, 1962 and 1966

	1962	*1966*
Index of domestic sales[a]	100	181
	Percentage of Net Sales	
Personnel costs, including travel		
Sales (mainly blockmen)	3.2	2.6
Technical	1.4	1.1
Repair parts	0.9	0.7
Wholegoods	0.7	0.6
Administration	0.9	0.7
Total personnel costs	7.1	5.7
Occupancy costs	1.0	0.9
Miscellaneous (postage, telephone, stationery)	0.5	0.4
Total branch accounted operating costs	8.6	7.0
Imputed investment costs		
Wholegoods and repair parts inventories and dealer receivables	6.0	5.0
Total branch office costs	14.6	12.0

[a]Data are for eight companies.
Source: Barber, *Report*, p. 178.

the figures. The two largest categories of expense, "sales" costs and imputed inventory and receivables cost (at 7.5 percent), are also the most suspect. The former category's largest component is expenditure on blockmen who visit each dealer between once a week and once a month, spending time "working up deals"[59] in addition to giving technical and business management advice and making sure that receipts from interest-free machines sold are promptly passed on to the company.

Of greater importance than "sales" costs are imputed investment costs, the lion's share of which are comprised of the dealer floor-planning expense of new and used equipment. It has been argued that to the unknown, but undoubtedly substantial, extent that this cost reflects the holding of tractors, it is certainly excessive. Furthermore, there is some physical deterioration when machinery at dealerships is stored out of doors, as it often is.[60] For this reason and because local or regional crop conditions generate demand which varies far more than the national average, local dealerships, where over half the unsold tractors are held at any one time, are inefficient places of storage. Indeed, it is sometimes necessary to take floor-planned machinery back unsold.[61]

The two categories of cost just examined accounted for 9.2 percent of total sales in 1962 and 7.6 percent in 1966.[62] Although it might be possible to achieve considerable savings in some of the subcategories reflected therein, particularly in supervision expense and in the financing of tractors, the production of tractors from new plants of efficient size and, or in additon to, production at some locations outside of North America offer the possibility of much greater real resource savings, and if such production were to result in lower prices, imputed investment costs would be cut pari passu.

To argue that the real resource costs of industry distribution practices are modest relative to possible production cost savings should not obscure either their considerable size nor their formidable role in maintaining entry barriers. Suitable blockmen, who like the good dealers they complement or compensate for the lack of, are a pivotally important advantage to established firms, and, as Table 3-6 demonstrates, the funds necessary for floor-planning greatly increase capital requirements.[63]

While it may well be that the greater geographical compression of all non-North American markets except Australia render the potential importance of floor-planning and close company supervision as expensive distributional devices less important than in the originating countries, there is little doubt that they are marketing innovations which, while certainly a threat to firms not employing them, are both wasteful and barrier-building. Their continued employment in North America and their growth outside of that market should be weighed negatively in evaluating the world industry's performance.

Chapter Thirteen

Public Policy

It has been determined that the structural characteristics of the world industry producing tractors have allowed for the survival of firms of greatly varying size with only a minor proportion of world production coming from plants of close to minimum efficient size and very different price-cost margins for firms operating from the same production area. Furthermore, international prices and costs have also been found to have been dramatically different and the former not fully explained by official protection or differences in distribution expense. Finally, much of the large distribution expense which characterizes the North American system appears to be essentially undesirable promotional expenditure.

THE POSITION OF AGRICULTURE VIS-À-VIS THE TRACTOR INDUSTRY

Before examining public policy proposals which could have the effect of stimulating price and cost adjustments in the industry, one is faced with the anterior question of explaining how the situation to 1970 developed. Specifically, why did farmers' groups in countries with relatively high prices not take action, either to get the prices lowered or even to demand explanations of why the prices were so high?

In the U.K. the situation has been subject to continuous monitoring by the machinery committees of the N.F.U. At least until 1970, however, the companies could always insist that, although prices might be rising, they were nevertheless the lowest in the Western world. It was not until the price explosion of 1970-71 that serious interest in abolishing tractor and machinery tariffs developed.

On the Continent, there is no evidence of concerted attempts to get rid of the tariffs in effect for farm machinery prior to a common external tariff, nor is there evidence that agriculture fought the final external tariff of 18 percent.

The argument that leadership circles in Continental agriculture would have seen little point in pressing for lower machinery prices because they were subject to joint determination with output prices in a highly protected market is not plausible, because many elements of a common agricultural policy in the EEC had been established by the late sixties while in 1970 there was great intra-EEC variation in tractor prices. Simple ignorance appears to be the explanation. This study found no evidence of Continental farmer interest in, or knowledge of, machine prices outside his own country, and farmer machinery boycotts seem to have been used more as a way of dramatizing the price-cost squeeze to politicians and the public rather than in an attempt to obtain leverage with the machinery makers.

The Australians seem never to have recognized officially that although no tariff is placed on machines coming from Britain, its major source of supply, untaxed entry of tractors does not assure prices close to those prevailing in the U.K. Farmers have engaged in the commercial policy-making process by opposing comprehensive tariffs and favoring manufacturing subsidies (in the early period even these were viewed with some skepticism). This attention was apparently judged to be sufficient, and no Australian study of tractor prices vis-à-vis the U.K. is known to the writer.

The largest and most persistent differences in prices have been the ones prevailing between the U.K. and North America. Official Canadian recognition of the difference goes back at least as far as 1960 and the Standing Committee on Agriculture and Colonization. Since that time Canadian farm groups have pursued many strategies for lowering the prices of tractors and other farm machines: more ambitious cooperative distribution of foreign machinery, "bootleg" imports, and the creation of the Royal Commission on Farm Machinery.

Why has no similar reaction been experienced in the United States? Much of the explanation may lie in the fact that, by contrast with his Canadian counterpart, the American farmer has for several decades had an opportunity to concentrate his attention on government subsidy in output prices, and this has been the focus of the efforts by the major farm organizations. As the distinguished agricultural economist, Varden Fuller, has suggested in correspondence with the writer,

Farm organizations did not find their own machinery manufacturing and related efforts to be profitable; they only got limited satisfaction from railroad, bank, and anti-trust legislation; prices received for well-known products as compared with prices paid by consumers are less obscure than machinery prices and hence have greater capacity for formulation as a politically maneuverable variable; and finally came PARITY, an ideological invention I believe is characteristic only of the United States; the image conveyed by parity always is, I believe, that costs must be accepted and accommodation should come in prices received.

In my view the concept of parity and its use in price supports embodies both the belief by U.S. farmers that they can obtain more government help on the output than on the input side and also an implication of moral responsibility on the part of government. Consider also that if government is going to help farmers get more income than free markets would bring them, impassive consumer powerlessness is no match for concentrated industrial power.[1]

In Canada, information and consideration of alternative courses of action seem to be centered in the unofficial provincial agricultural federations, but in the U.S. neither the Grange, Farm Bureau, the National Farmers Union, nor the National Farmers Organization has engendered or entertained any discussion of the subject as far as the writer is aware, and the rural press in the United States is certainly culpable for failing to report even so much as the allegations being made in Canada. It is a virtually inescapable conclusion that the rather substantial amount of farm machinery advertising in these media played a role in their silence.

The closest any official investigation ever got to identifying the problem in the U.S. by the early seventies was the House Committee on Agriculture's "Farm Cost-Price Squeeze" Hearings of 1961. Although the hearings discovered the difference between the U.S. and the U.K. prices and manufacturing costs, and committee members were aware of the no-tariff situation, the obvious inferences were never drawn. This may have been due in part to the divided loyalties of some of the members of the Committee when questioning company and union spokesmen. The most knowledgeable questioner on the subject was a Congressman from Illinois where both Deere and International Harvester are headquartered (although neither in his district).[2]

In the absence of a very substantial increase in interest on the part of U.S. farmers, however generated, not only will their own prices continue to be largely independent of the rest of the world, but so, too, to a great extent will those in Canada. Certainly, no major concessions to Canada alone are thought practicable by the large companies because, however ignorant is the U.S. farmer, arbitrage of obviously identical products along the long open border would be inevitable, and an attempt to thwart it by tying warranty work to a specific dealer could be expected to produce outrage. CCIL's advance payments scheme and the various attempts by both Eastern Europeans and official bodies in Canada to gain acceptance by farmers for Rumanian, Polish, and Russian machines may put some downward pressure on prices but not to the extent possible if private and public groups in the much larger U.S. market were similarly interested.

One favorable development in the U.S. would be direct foreign purchase, perhaps from Eastern Europe, by large industrial users or public agencies. The performance of the equipment, presuming it would be comparable to that presently being used in agriculture, could "soften up" the agricultural market.

Unfortunately, as late as the early seventies, this possibility was largely blocked because only Poland of the Eastern countries enjoyed most-favored-nation trading status. Tractors from other Eastern countries destined for non-agricultural use would have faced the 27.5 percent duty which prevailed in 1930. Removal of the penalty for non-agricultural sales would make possible efforts by Eastern sellers to offer machines for all uses at uniformly low prices. The 5.5 duty on other tractors imported for non-farm use which remained after the Kennedy Round should also be abolished.

In the enlarged EEC one can perhaps safely predict a final common price level as agriculture becomes more unified and information is more widely sought and shared. This means that the prices of Britain and Italy will ultimately meld with those of France and Germany at some intermediate level.[3] This movement can be expected to be facilitated by the almost certain illegality under the Treaty of Rome of export-banning agreements of the kind discussed in Chapter Ten. The contracts appear "likely to affect trade among the Member States and have as their object or result the prevention, restriction or distortion of competition within the Common Market . . ." (Article 85-1).[4] Any scheme accomplishing the same end, such as implicit dealer quotas, would probably be equally illegal. It is also likely that any sustained increase in tractor prices in Britain would bring the N.F.U. to the forefront of Common Market agricultural efforts to get rid of the 18 percent common external tariff completely, however the overall agricultural policy of the EEC develops.

A considerable rift was caused in the European farm machinery manufacturers' association in the early sixties, when non-Italian members suggested that the financial support extended to Italian farmers should be expanded to include the purchase of foreign-made machinery, and that the "Fanfani law" in fact contravened both the General Agreement on Tariffs and Trade and the Treaty of Rome. The response of Italian manufacturers was to reject the proposal as "prejudicial to the legitimate position of the Italian industry" because the law "has for its ultimate object the finding of increased employment for Italian workers."[5] In the early seventies, it appeared that Italy still discriminated in essentially an unchanged way, but British membership in the EEC will undoubtedly force official attention to the practice, and it should certainly be resolved against the Italian policy. Increased competition in the domestic market might prod the leading Italian firms, Fiat and Same, to consider foreign marketing more seriously. (Same's passive role in most foreign markets through the use of White as a marketing agent might be reconsidered, and the North American market was not covered by their agreement.)

In Australia, farmer ignorance and lack of interest would seem capable of perpetuating the present situation indefinitely with the bounty payments in the seventies perhaps partly remitted to Deere in the United States. Indirect amelioration in Australia would appear to call for the same solution as was suggested for the United States, and could come about more easily because

Zetor, the most impressive Eastern line, was being actively sold in Australia by the late sixties.

PUBLIC POLICY

It is against this background that possible changes in national public policy must be considered. For Canada, several measures were advanced by the RCFM in its *Report*, where it is suggested that the proposals "would clearly be more effective if the U.S. government were to adopt similar measures."[6]

The most important recommendation was that "the government should prohibit the floor-planning of new and used farm machines on an interest-free basis in the hands of dealers. To be effective, this measure would need to be supported by a ban on consignment selling, and a provision for minimum interest rates on sales to dealers on a credit basis."[7]

The declared objective of the ban is twofold. First, it is intended to make entry by new competitors easier (presumably by lowering both the absolute cost and product differentiation barriers to entry) and secondly, it is aimed at the realization of resource savings by an assumed reduction of inventory in the hands of dealers. This is a measure particularly commended to U.S. authorities for consideration.

The proposal can be examined on the grounds of both legality and efficacy. Whatever the status of such ad hoc banning of commercial practices under Canadian law, it would seem unlikely that floor-planning could be attacked in the U.S. The practice violates none of the antitrust laws, and there is no way wasteful but barrier-building sales promotion can be tackled any more easily in this case than elsewhere in the economy. Indeed, one would have thought that the offense here is more subtle than extremely large advertising expenditure as a percentage of industry sales, a practice which remains to be successfully attacked in U.S. courts.[8]

A further criticism would apply even to the proposal for Canada alone. Although consignment selling might be banned, it is not at all clear that the Canadian authorities could prevent forward integration if the companies found the massive retail inventories to be central to their marketing success. The development of an adequate incentive scheme for managers would present a challenging problem but very probably not an insoluble one.

Another suggestion is that official publicity be given to the relative cheapness of the government-guaranteed controlled interest loans mentioned in Chapter Two.[9] The *Report*, however, is vague about whether or not anything should be done to make the loans more attractive to lenders. In any event, any elements of subsidy for tractor purchase need justification, and the *Report* offers none except the implied diminution of entry barriers. It is not clear how important retail financing is as a barrier to entry, however, and a potential entrant could presumably arrange for retail sales finance with an independent company as

White Motor did. It should also be noted that from a societal point of view there is no obvious rationale for the effective and substantial subsidies for tractor purchase offered by the British and Italian governments.

The RCFM recommends that exclusive dealing be completely banned;[10] this is a policy avenue which might be worth considering in both the United States and Canada. Although previous U.S. court reluctance to disallow "adequate attention" to one line might not stand up under a full examination of rural entrepreneurial scarcity, additional federal legislation to define more clearly where exclusive dealing should be permitted might be necessary. Such a policy also deserves a very close look in Australia where the problem may be even more severe. In other markets, exclusive dealing appears to be of less critical importance, and the manufacturers' insistence that his products be adequately represented has less to be weighed against it, although undoubtedly it has a significant anti-competitive effect everywhere.

In the *Special Report on Prices* of 1969, a "reverse dumping duty" was proposed for Canada [11] which was aimed at price discrimination against the Canadian farmer. The RCFM apparently abandoned this idea between 1969 and 1971 because it is not suggested in the *Report*, although no explanation is given for its withdrawal. The proposal as presented is fraught with ambiguity and error, but it is the writer's opinion that the reverse dumping duty would be an impractical plan which in most situations (even when properly formulated) would have the same impact as a simple ban on discriminatory pricing.

If the RCFM proposals do not seem very promising as solutions, is there any recourse in North America to accomplish lowered entry barriers, a reduction in price discrimination, or cheaper distribution?

One kind of lowered entry barrier could result from better information in the hands of the U.S. farmer. If the farmer could be made aware by his own organizations of the persistent and substantial discrimination practiced against him in the past, it is not unreasonable to assume that he would become at least somewhat annoyed (his Canadian counterparts certainly did!), and the trust which is such a large part of the product differentiation barrier carefully nurtured by the companies over the years would be diminished thereby. A demonstration that fully comparable tractors have been sold for up to two decades in other parts of the world for vastly lower prices than offered to the U.S. farmer should help convince him that a bargain tractor price is something to be carefully considered and not necessarily a sign of inferior design or manufacture. The result, if combined with sufficient information about the qualities of potentially entering machinery, might favorably dispose buyers to the blandishments of a newcomer.

Does the Robinson-Patman Act, supposedly aimed at price discrimination, offer any remedy? It is possible that an attack based upon the deleterious impact on certain segments of American agriculture in international markets from producer durable discrimination might be considered, although damages might be difficult to demonstrate. Nevertheless, international price discrimination in producer goods, when competition in final product markets is in any sense international, seems to constitute a violation in the sense of a "secondary line" injury under Section 2(a): that an injury accrues to those competing with buyers to whom discriminating (lower) prices have been granted. Although Section 2(b) of the law as presently interpreted, makes "meeting competition" an adequate ground for a firm not to charge uniform prices in different markets, [12] the latter defense appears not to have been tested where competition farther "downstream" is affected.

A more direct avenue of attack might be through Section 1 of the Sherman Act because it appears that the restriction of British sales for use only in the U.K. by branches of North American firms and the tying of warranties to specific dealers have in fact served to "restrain trade" in a way detrimental to the U.S. purchaser. Close scrutiny should also be given to the licensing or distribution arrangements made by U.S. firms with potential foreign competitors.

The policies just discussed would be worth exploring if U.S. agriculture ever develops an interest in the issue, for they offer the possibility of ending discriminatory practice, although success would raise prices abroad as well as lowering them in North America.[13]

Other developments would be likely if discrimination were forced to end. Officials of the major companies have stressed the higher cost of distribution in North America (and one infers Australia) than elsewhere. If North American farmers could order "F.O.B. Coventry" or Basildon, not only would any genuine discrimination be overcome, but the extra costs of North American distribution would have to be given a close look. Floor-planning might be reduced or even abolished with a resultant diminution in the value to the manufacturers of the smaller dealerships which would presumably drastically cut their display of wholegoods in the face of reduced manufacturer support. Much of their function would thus be eliminated and their demise hastened. The international differences in labor and parts stocking costs might dictate a shorter warranty period or even optional warranty protection. In short, if discrimination were banned, the companies would be forced to give careful attention to non-machine costs, and any that could not be presented to farmers as charges would, of course, constitute some discrimination *in favor* of North America.[14]

What is being suggested as a goal for North American farmers bears some resemblance to the choice International Harvester claimed for its overseas customers over a decade ago.

> There is no allocation of markets to foreign subsidiaries, . . . and in any country the customer can choose himself whether to buy the product of a U.S. plant, or a plant in Britain or France or wherever production is maintained.[15]

The catch then, however, was that customers in different countries were quoted vastly different F.O.B. prices for the same equipment.[16]

A minor modification of U.S. public policy which might make floor-planning somewhat less attractive and would increase purchaser understanding would be "sticker pricing" of the kind mandatory in the automobile industry, and which Deere began voluntarily in 1969. The features should probably be more detailed than those presented by Deere, however, and if the scheme were adopted in all countries it would make possible accurate international price comparisons.

Public policy revision in the EEC should include the elimination of the 18 percent external duty, despite the fact that Common Market farmers will soon have access to the products of both low-cost national industries—those of Italy and Britain—because maximum access to the EEC for Eastern European and Japanese machinery should still be a desideratum.

Our investigation of cooperatives found neither the exercise of monopsony power nor general attention to the question of farm machinery prices. Nevertheless, however questionable their efficacy for many economic purposes,[17] cooperatives could certainly perform the function of purchasing agents to bring machinery from, say, Italy to Germany, hence hastening price unification, and one would have thought that the cooperative distributors of Fiat equipment in Italy would be poorly placed, both morally and legally, to avoid giving assistance to their EEC counterparts. Such action on any scale, of course, would await a recognition of the problem by farmers themselves. In any event, it would seem wise for cooperatives and other groups to discourage the growth of floor-planning and press for distribution channels which move equipment from factory to farmer as directly as possible.

Recommendations for change in Australian public policy can be stated with confidence. According to the official position of the Australian Tariff Board in 1955, as expressed by its chairman,

> . . . a condition precedent to protection would be the sound prospect of success of the developing of the industry, i.e., the ability of the industry to produce economically for commercial purposes.[18]

The Board appears to have believed that in granting support for local tractor manufacture, it was only aiding Chamberlain in an essential way, and that International Harvester would have manufactured tractors in Australia in any case.

> And while it is comforting to know that there is in Australia an engineering industry capable of producing an intricate piece of machinery without protection or assistance (International Harvester), an industry setting a standard for other industries—the Board takes the sensible view that an industry or a unit of an industry is not necessarily inefficient because it fails to reach that high standard.[19]

As previously noted, whether International Harvester could or could not have profitably produced tractors for sale in Australia in 1955, it was clearly finding it increasingly more profitable *not* to do so, a trend which only the bounty increase of 1959 temporarily turned around. The present or projected volume of Australian total sales clearly does not warrant even one small tractor plant, and almost any projection of Australian efficiency wage rates in international prices finds it a high absolute cost country which would benefit from being a tractor importer rather than a producer. The bounty system should therefore be abandoned.[20]

Like his North American counterpart, the Australian farmer would seem to be well served by a policy which would allow him to purchase F.O.B. point of manufacture, and the Australian government might attempt to accomplish such a policy in cooperation with the North Americans. The implications for Australian distribution appear similar to those for North America.

HEALTH AND SAFETY REGULATIONS

The increased public concern in many countries in the sixties for the health and safety of tractor users appears to be generally appropriate. The argument of user ignorance of the risks faced can be applied to mandatory requirements that affect only the farmer himself or perhaps his family. The alternative of providing detailed information on risks and allowing for individual choice might not be superior given the government's role in every country as provider of last resort. These arguments also apply when the machinery is to be operated by employees and where additionally there is a long-standing tradition in virtually all countries of not allowing variation in market wages to carry the entire burden of adjusting for danger to the employee.

Even if practically all circumstances argue for the legitimacy of government concern, the weighing of the costs and benefits of different devices with different degrees of effectiveness remains an enormous problem. Although there are many important sources of tractor-related injuries and fatalities, neither the extent of the problem nor the effectiveness of various remedies has been established for any area except tractor turnovers, and even here the information is very rudimentary and confined mainly to fatalities. On the basis of the fatality information alone, however, a rather strong case can be made for compulsory rollbars on not implausible assumptions. Assuming that unit tractor demand does not change significantly, and using the 1970 price of $200 per rollbar, the annual cost of the compulsory scheme in the United States would be about $40 million. Assuming no "retrofit" requirement, it would be about 15 years before nearly the entire tractor stock of the country was so fitted. On the undoubtedly somewhat exaggerated assumption that virtually all fatalities from tractor overturns would then be eliminated and using the estimate of fatalities of 600 (although this was a high estimate for 1970, increasing field speeds might raise it in later years without protection), this would imply a continuing annual cost of $67,000 per life saved. Only a few years of production by most farmers, discounted at any reasonable rate, would be enough to balance this cost.[21]

Proposals for the study of rural accidents, such as those made in 1971 by the U.S. Department of Transportation, appear to be an important step in the right direction.[22] International cooperation in the investigation of different types of accidents could presumably diminish duplication of effort, increase the confidence with which conclusions about the causes of accidents could be held, or some combination of the two.

PARTS STANDARDIZATION

Another area on which government attention might be focused is parts standardization; it was explained in Chapter Ten that the major firms will probably never voluntarily achieve interbrand interchangeability to any substantial extent. The Canadian *Report* suggested that an official agency be established to determine the practical limits to interchangeability, to advertise its importance, and to publicize the extent to which machines of various makes meet interchangeability standards.[23] Such an agency might be useful for other countries as well, and, as in the case of health and safety research, a common international effort could avoid duplication.

CONCLUDING COMMENT

Agriculture as a purchaser of capital equipment has been highly fragmented in most countries and unaware of real or potential alternative sources of supply. In this setting it is not surprising that the multinational corporations have been able to treat national markets largely in isolation, virtually as if sufficient official barriers existed even when they did not. A posture of indignation toward the companies is both futile and inappropriate. As an editorial commenting on the RCFM accusations of discrimination in the British *Agricultural Machinery Journal* put it rather baldly,

> In a free economy, price depends on the balance between supply and demand and should not depend on cost. . . . Manufacturers consider they have a duty to maximize their profits.[24]

Although the point is not expressed with technical accuracy, its essential truth remains. The only way any national agriculture can be certain of avoiding discrimination is through vigilance and the support of appropriate public policy, and this effort is in conflict with the interests of those offered relatively low prices under discrimination.[25]

When one addresses the broader problem of the efficient production of tractors and their sale at the lowest possible price from the standpoint of world agriculture as a whole, solutions do not abound. The irreducible problem is that a rather high level of concentration is essential for the realization of scale economies.[26] This is very important because, as Oliver Williamson has pointed out in the context of merger policy,[27] if income distribution and certain other considerations are set aside, it takes a modest increase in the realization of scale economies to outweigh in efficiency terms a rather substantial diminution in competitiveness measured as an increase in price. Where prior to a merger there is no market power (such market power makes the following trade-off somewhat less dramatic), a gain in average costs of only 2 percent is sufficient to compensate for a price increase of 20 percent when demand elasticity is -1.0; compensation of 4.4 percent is necessary when the elasticity is -2.0. In this writer's judgment a policy of attempting to increase competitiveness in the indusry through enforced deconcentration would be a very doubtful course given the shapes of the estimated cost functions. Dividing Ford and Massey-Ferguson up into only two firms apiece would raise their costs from 10 to 15 percent and would need to be compensated for by drastic price reductions to make the policy worthwhile in efficiency terms, assuming that the input price schedule remained the same. In fact, it appears that the strongest case for deconcentration on efficiency grounds could be made if it could be shown that such fragmentation would lead to a more complete and rapid migration of the industry to areas of lower absolute cost—a change in the relevant cost functions for the producing units under consideration rather than a movement along them. Such a development, however, is speculative in the extreme. Furthermore, it is highly doubtful that such a diminution of the size of the largest firms in the industry through public policy measures is feasible even if it were otherwise promising.

What developments appear most propitious for an improvement in the world industry's performance? It seems clear that an increase in the size of Massey-Ferguson, Ford, or International Harvester through acquisition is antithetical to competition and should be prevented. This, however, is only a meager and passive policy which might allow other firms to be driven into oblivion and the gradual emergence of a world industry structure even less promising than the present one. The encouragement of the growth of several more firms to the approximate size of the present industry leaders and a balanced participation of all major firms in markets around the world appears to be the best possible structure consonant with an efficient volume of production by each firm. This development assumes the demise of sub-optimal firms and also appears to offer some hope for the degree of price competition necessary to induce the world industry to source from areas of low absolute cost. Even if such an unlikely outcome were to develop, however, the level of concentration would still be in Bain's "High-Moderate" range, and, coupled with high barriers to entry, this level of concentration could well produce unsatisfactory results.

Better performance would be more completely assured for any size distribution of sellers if the tractor producers faced strong buyers rather than the atomistic market they face at present,[28] but attaining the monopsony power which could assure minimal price-cost margins and lowest real cost of supply seems an elusive ideal. Short of that, maximum impact could presumably be attained if interested groups in many countries coordinated their activities to gather information and to examine all available sourcing options worldwide. Even speculation about such a development seems premature, however, in light of a nearly general failure to recognize the problem at the national level.

Selected Tractor Price Comparisons

Where possible the price per horsepower figures presented for each country in Figure 10-1 are based on the average of two comparable models within the relevant size range for each year of observation. Net wholesale price (NWP) was derived from Suggested Retail Price (SRP) by applying the dealer discounts prevailing in the mid-to-late sixties (Barber, *Prices,* p. 100). Available information suggests that they were very similar (within a couple of percentage points) in all countries in previous years. All prices are expressed in U.S. dollars at prevailing exchange rates. All power figures are given at either tested or estimated power take-off horsepower. All tractors are diesel-powered units, with hydraulics, and the equipment is as closely matched both intra- and internationally as available information allows.

North America

Year	Model	Power	U.S. Dollars SRP	U.S. Dollars NWP	$/h.p.	Average $/h.p.
			30–40 h.p.			
1955	Ford FMD–12	38	2665	1950	51	53
	Oliver Super 55	34	2564	1875	55	
1960	M–F TO–35 (Dual Clutch Standard)	34	3035	2216	65	64
	Ford Dexta	31	2640	1927	62	
1966	M–F 135 Deluxe Diesel	38	3478	2540	67	67
	Ford 2000	36	3308	2415	67	
1970	M–F 135 Standard 6-Speed	38	4119	3007	79	82
	Ford 2000 8-Speed	31	3793	2608	84	
			45–55 h.p.			
1955	IH "Farmall" 400	47	3630	2650	56	52
	Oliver Super 88	55	3644	2660	48	
1960	IH Diesel Utility	52	3957	2888	56	57
	Oliver 770 Wheatland	52	4153	3032	58	
1966	M–F 165 Diesel Standard	52	4847	3538	68	69
	Ford 5000	48	4554	3324	69	
1970	M–F 165 Diesel Standard	52	5575	4069	78	75
	Ford 4000 8-Speed	53	5217	3808	72	

United Kingdom

Year	Model	Power	U.S. Dollars		$/h.p.	Average $/h.p.
			SRP	NWP		
		30–40 h.p.				
1955	Fordson Major Diesel	38	1613	1323	35	41
	David Brown 30D	32	1789	1467	46	
1960	IH B–275	33	1728	1417	43	42
	M–F 35	34	1641	1346	40	
1965	M–F 135	38	2170	1779	47	47
	Ford 2000	36	2016	1653	46	
1970	M–F 135	38	2254	1848	49	51
	Ford 2000	31	1973	1618	52	
		45–55 h.p.				
1955	David Brown 50D	47	2709	2221	47	44
	IH Farmall Super BMD	47	2276	1866	40	
1960	Nuffield Universal 4DM	52	1954	1602	31	34
	M–F 65	47	2086	1711	36	
1965	M–F 165	52	2562	2101	40	41
	Ford Super Major 5000	56	2870	2353	42	
1970	M–F 165	52	2580	2115	41	41
	Ford 4000 8-Speed	53	2662	2183	41	

France

Year	Model	Power	U.S. Dollars SRP	U.S. Dollars NWP	$/h.p.	Average $/h.p.
			30–40 h.p.			
1956	Renault D35	33	3466	2911	88	90
	Someca DA 50L	36	3921	3294	92	
1960	Renault N71	33	2470	2080	63	69
	M–F 835	31	2761	2319	75	
1966	Renault Super 5D	35	2990	2512	72	71
	M–F 135	38	3140	2638	69	
1970	Renault 55	35	2858	2323	69	65
	M–F 135	38	2765	2401	61	
			45–55 h.p.			
1956	Vendeuvre 66	53	6363	5345	101	95
	Sift TL4	56	5906	4961	89	
1960	M–F 865	52	3750	3150	61	61
1966	M–F 165	55	4120	3461	63	65
	Renault Master II	53	4240	3562	67	
1970	M–F 165 MK III	56	3699	3107	55	59
	Renault Master II	53	3893	3270	62	

Italy

Year	Model	Power	U.S. Dollars		$/h.p.	Average $/h.p.
			SRP	NWP		
			30–40 h.p.			
1955	OM					
	35/40 R	36	3024	2389	66	72
	Same					
	DA 38	35	3424	2704	77	
1960	Fiat					
	411R	37	2600	2054	56	56
1964	Fiat					
	411R	37	2600	2054	56	56
1970	Fiat					
	400	37	2728	2155	58	58
			45–55 h.p.			
1955	Fiat					
	55R	51	5104	4032	79	75
	Same					
	DA55	51	4560	3602	71	
1960	OM					
	512R	56	3168	2502	45	45
1966	Fiat					
	615	54	3473	2744	50	50
1970	Fiat					
	550	52	3457	2731	53	53

Germany

Year	Model	Power	U.S. Dollars		$/h.p.	Average $/h.p.
			SRP	NWP		
		30–40 h.p.				
1960	Hanomag R435	35	2960	1924	55	56
	IH D436	36	3152	2049	57	
1967	Deutz D–4005	35	4139	2690	77	77
1970	Deere 1020	39	–	2781	71	71
		45–55 h.p.				
1960	Deutz F3L 5147 NKF	50	4214	2739	55	53
	Porsche Master	55	4275	2779	55	
1967	Deutz 5505	52	5203	3381	65	65
1970	Deere 2020	54	–	3568	66	66

Australia

Year	Model	Power	U.S. Dollars SRP	NWP	$/h.p.	Average $/h.p.
			30–40 h.p.			
1955	Fordson Major	38	2583	2092	55	55
1960	Fiat 411R	37	2397	2173	52	55
	Ford Dexta	31	1941	1760	57	
1965	Fiat 411R	37	2529	2048	55	55
1970	IH 434	36	3130	2504	70	70
			45–55 h.p.			
1959	Ferguson 65	49	3125	2532	52	52
1961	Chamberlain Champion 96	58	3560	2812	48	48
1965	Nuffield 10/60	55	3290	2665	48	49
	Chamberlain Canelander MK III	58	3573	2894	50	
1970	IH Farmall 564	54	3936	3188	59	59

Source: See material cited for Tables 10–1 through 10–4. In addition, price information was obtained from *Tracteurs et Machinisme Agricole, Motorisation Agricole, Informatore Agrario, Macchine e Motore Agricoli*, and the files of the Royal Commission on Farm Machinery.

Bibliography

BOOKS AND PAMPHLETS

Annual Reports of Allis-Chalmers, J.I. Case, Deere, Ford, International Harvester, Massey-Ferguson, White (and predecessor companies), various annual issues.

Agricultural Machinery and Tractor Dealers' Association. *National Survey of Trading Costs, Margins, and Profits.* Penn Place, Rickmansworth, Herts.: Agricultural Machinery and Tractor Dealers' Association, various annual issues.

Bain, Joe S. *Barriers to New Competition.* Cambridge: Harvard University Press, 1956.

Bain, Joe S. *Industrial Organization.* New York: John Wiley & Sons, Inc., 1968.

Baldwin, Robert E. *Non-Tariff Distortions of International Trade.* Washington: The Brookings Institution, 1970.

Baum, Warren. *The French Economy and the State.* Rand Corporation Research Study. Princeton: Princeton University Press, 1958.

Beerman's *Financial Yearbook of Europe,* various annual issues.

Behrman, J.N. *Some Patterns in the Rise of the Multinational Enterprise.* Chapel Hill: Graduate School of Business, University of North Carolina, 1969.

Brash, D.T. *American Investment in Australian Industry.* Cambridge: Harvard University Press, 1966.

Caves, Richard E. *Air Transport and Its Regulators.* Cambridge: Harvard University Press, 1962.

Caves, Richard E. *American Industry: Structure, Conduct, Performance.* 3rd ed. Englewood Cliffs, New Jersey: Prentice-Hall, Inc., 1972.

Caves, Richard E., and Jones, Ronald W. *World Trade and Payments:An Introduction.* Boston: Little, Brown and Company, 1973.

Conant, Michael. "Aspects of Monopoly and Price Policies in the Farm Machinery Industry Since 1902." Unpublished Ph.D. dissertation, University of Chicago, 1949.

Cromarty, W.A. *The Demand for Farm Machinery and Tractors.* East Lansing: Michigan State University Press, 1959.

Dirlam, Joel B., and Kahn, Alfred E. *Fair Competition: The Law and Economics of Antitrust Policy.* Ithaca, New York: Cornell University Press, 1954.

Donaldson, Lufkin and Jenrette, Inc. *The European Agricultural Equipment Industry and Competitive Positions of North American Producers.* New York: Donaldson, Lufkin and Jenrette, Inc., 1966. (Mimeographed.)

Dun and Bradstreet, Inc. *Key Business Ratios.* New York: Dun and Bradstreet, Inc., various issues.

Dunning, John H. *American Investment in British Manufacturing Industry.* London: George Allen & Unwin Ltd, 1958.

Eckles, Elvis L. "The Development of Oligopoly in the Farm Implement Industry." Unpublished Ph.D. dissertation, University of Illinois, 1953.

Farm and Industrial Equipment Institute. *Farm and Industrial Equipment Facts.* 4th ed. Chicago: Farm and Industrial Equipment Institute, various annual issues.

Farm and Power Equipment. *Why Do Farmers Buy Their Machines Where They Do?* St. Louis: National Farm and Power Equipment Dealers' Association, 1965.

Farm Equipment Institute. *Facts, 1959 Edition.* Chicago: The Farm Equipment Institute, 1959.

Fellner, William. *Competition Among the Few; Oligopoly and Similar Market Structures.* New York: Augustus M. Kelley, 1965.

Galbraith, John Kenneth. *American Capitalism; The Concept of Countervailing Power.* Boston: Houghton Mifflin Company, 1952.

Gray, R.B. *Development of the Agricultural Tractor in the United States, Part I.* St. Joseph, Michigan: American Society of Agricultural Engineers, 1956.

Gray, R.B. *Development of the Agricultural Tractor in the United States, Part II.* St. Joseph, Michigan: American Society of Agricultural Engineers, 1956.

Harvard Business School. *Deere and Company.* ICH 13G 124R BP 865R, Cambridge: President and Fellows of Harvard College, 1966.

Harvard Business School. *A Note on the Farm Equipment Industry.* ICH 13G 122 BP 864R2. Cambridge: President and Fellows of Harvard College, 1968.

Heady, Earl O., and Tweeten, Luther G. *Resource Demand and Structure of the Agricultural Industry.* Ames: Iowa State University Press, 1963.

Hewitt, William A. "Remarks before the New York Society of Security Analysts." Deere and Company, February 14, 1957. (Privately Printed.)

Implement and Tractor Redbook, various annual issues.

Johnson, D. Gale. *Farm Commodity Programs: An Opportunity for Change.* Washington: American Enterprise Institute for Public Policy Research, 1973.

Kaplan, A.D.H.; Dirlam, Joel B.; and Lanzillotti, Robert F. *Pricing in Big Business.* New York: The Brookings Institution, 1958.

Linder, Staffan Burenstam. *An Essay on Trade and Transformation.* New York: John Wiley & Sons, Inc., 1961.

Linneman, R.E. "The United States Tractor Industry in Selected Foreign Markets." Unpublished Ph.D. dissertation, University of Illinois, 1964.

Loucks, William N., and Hoot, J. Weldon. *Comparative Economic Systems.* 4th ed. New York: Harper & Brothers, 1952.

MacDonald, Neil B. "Farm Machinery Costs." Paper presented before the Edmonton Chamber of Commerce, November 13, 1970. (Mimeographed.)

Martel, Phillip A. *The 1965 Ford Tractor Engine Family.* New York: Society of Automotive Engineers, Inc., 1965.

Massey-Ferguson Limited. "Statement on the Special Report on Prices of the Royal Commission on Farm Machinery." Toronto: March 3, 1970. (Mimeographed.)

Moody's *Industrial Manual,* various annual editions.

National Farm and Power Equipment Dealers' Association. *Cost of Doing Business Survey.* St. Louis: National Farm and Power Equipment Dealers' Association, various annual issues.

National Farm and Power Equipment Dealers' Association. *Official Guide.* St. Louis: National Farm and Power Equipment Dealers' Association, various semi-annual issues.

Neufeld, E.P. *A Global Corporation: A History of the International Development of Massey-Ferguson Limited.* Toronto: University of Toronto Press, 1969.

Nevins, Alan, and Hill, Frank Ernest. *Ford: Decline and Rebirth, 1933-1962.* New York: Charles Scribner's Sons, 1963.

Nevins, Alan, and Hill, Frank Ernest. *Ford: Expansion and Challenge: 1915-1933.* New York: Charles Scribner's Sons, 1957.

Pashigian, B.P. *The Distribution of Automobiles, An Economic Analysis of the Franchise System.* Englewood Cliffs, New Jersey: Prentice-Hall, Inc., 1961.

Peterson, Clifford L. "Remarks before the American Industrial Development Council Conference." St. Louis: April 5, 1965. (Mimeographed.)

Phillips, W.G. *The Agricultural Implement Industry in Canada.* Toronto: University of Toronto Press, 1956.

Pitfield, MacKay, Ross and Company, Limited. *The Farm Machinery Industry.* Ottawa: Pitfield, MacKay, Ross and Company, Limited, 1969.

Political and Economic Planning. *Agricultural Machinery.* London: Political and Economic Planning, 1949.

Power Farming in Australia and New Zealand Technical Annual, various annual issues.

Rayner, A.J. *An Econometric Analysis of the Demand for Farm Tractors.* University of Manchester, Department of Agricultural Economics, Bulletin No. 113, October, 1966. (Mimeographed.)

Robinson, Joan. *The Economics of Imperfect Competition.* London: Macmillan & Co., 1933.

Schelling, Thomas C. *The Strategy of Conflict.* New York: Oxford University Press, 1960.

Scherer, F.M. *Industrial Market Structure and Economic Performance.* Chicago: Rand McNally & Company, 1970.

Schumpeter, Joseph A. *Capitalism, Socialism, and Democracy.* London: Unwin University Books, 1954.

Shearer, Warren Wright. "Competition Through Merger: An Economic Analysis of the Farm Machinery Industry." Unpublished Ph.D. dissertation, Harvard University, 1951.

Shubik, Martin. *Strategy and Market Structure.* New York: John Wiley & Sons, Inc., 1960.

Singer, Eugene M. *Antitrust Economics: Selected Legal Cases and Economic Models.* Englewood Cliffs, New Jersey: Prentice-Hall, Inc., 1968.

Straszheim, Mahlon R. *The International Airline Industry.* Washington: The Brookings Institution, 1969.

Syndicat des Constructeurs Français de Matériels de Motoculture. *L'Industrie des Tracteurs et des Motoculteurs.* Paris: Société Nouvelle Mercure, 1955.

Terre Moderne. Paris: Les Tracteurs Agricoles Renault, 1950.

Unione Nazionale Costruttori Macchine Agricole. *Trattrici.* Rome: Unione Nazionale Costruttori Macchine Agricole, 1969.

Utenti Motori Agricole. *La Meccanizzazione Agricola in Italia.* Rome: Utenti Motori Agricole, various annual issues.

White, Lawrence J. *The Automobile Industry Since 1945.* Cambridge: Harvard University Press, 1971.

Whitney, Simon N. *Antitrust Policies: American Experience in Twenty Industries,* Volume II, *Famous Antitrust Cases.* New York: The Twentieth Century Fund, 1958.

Wigglesworth, E.F., ed. *Readings in Policy and Practice in International Business.* New York: Thomas Ashwell & Company, Inc., 1961.

Wilkins, Mira and Hill, Frank Ernest. *American Business Abroad: Ford on Six Continents.* Detroit: Wayne State University Press, 1964.

Worthington, Wayne H. *50 Years of Tractor Development.* New York: Society of Automotive Engineers, 1966.

ARTICLES

Abrahamsen, Martin A. "Discussion: The Changing Structure of Markets for Farm Machinery." *Journal of Farm Economics,* 40 (December, 1958), pp. 1182–85.

"Back to the Land." *Forbes,* November 15, 1968, pp. 30–32.

Bernasek, M., and Kubinski, A.M. "Agricultural Machinery and Implements." In Alex Hunter, ed., *The Economics of Australian Industry.* Melbourne: Melbourne University Press, 1963, pp. 460–492.

Bishop, R.L. "Monopolistic Competition and Welfare Economics." R.E. Kuenne, ed., *Monopolistic Competition Theory: Studies in Impact.* New York: John Wiley & Sons, Inc., 1967.

Bookman, George. "Farm Machinery Shifts Gears." *Fortune,* July, 1961, pp. 129–132, 188, 193–194.

Calcatierra, E.; Mazzocchi, G.; Lombardini, S.; and Vito, F. "The Main Outlines of the Structure of the Italian Economy." In R. Frei, ed., *Economic Systems of the West.* Volume II. Tübingen: J.C.B. Mohr–Paul Siebeck, 1957, pp. 81–116.

"The Case of Bootleg Tractors." *Good Farming,* August, 1969, pp. 18–31.

Caves, Richard E. "Foreign Investment, Trade and Industrial Growth." The Royer Lectures. University of California, Berkeley. December 1-2, 1969. (Mimeographed.)

Conant, Michael. "Competition in the Farm Machinery Industry." *Journal of Business,* 26 (January, 1953), pp. 26-36.

"Converging Wheels" *Successo,* October, 1968, pp. 50-56.

Cowling, Keith, and Rayner, R.J. "Price, Quality, and Market Share." *Journal of Political Economy,* 78 (December, 1970), pp. 1292-1309.

"Deere & Co." *Forbes,* June 15, 1966, pp. 30-36.

"Der Deutsche Traktorenmarkt Gehört Jetzt den Groszen." *Frankfurter Allgemeine Zeitung,* February 2, 1970.

Dieffenbach, E.M., and Gray, R.B. "The Development of the Tractor." U.S. Department of Agriculture, *Yearbook of Agriculture, 1960.* Washington: U.S. Government Printing Office, 1960, pp. 25-45.

Edwards, Corwin D. "Size of Markets, Scale of Firms, and the Character of Competition." In E.A.G. Robinson, ed., *Economic Consequences of the Size of Nations.* London: Macmillan & Co. Ltd., 1960, pp. 117-130.

"Fiat l'Offensive." *Entreprise,* December, 1970, pp. 11-17.

"Fighting on Two Fronts." *Forbes,* October 1, 1968, pp. 19-20.

Garelli, France. " 'Power Plants' for Italian Agriculture." *Successo,* April, 1968, pp. 99-104.

Griliches, Zvi. "The Demand for a Durable Input: Farm Tractors in the United States, 1927-57." In A.C. Harberger, ed., *The Demand for Durable Goods.* Chicago: The University of Chicago Press, 1960, pp. 181-207.

Grubel, Herbert G. "Intra-Industry Specialization and the Pattern of Trade." *Canadian Journal of Economics and Political Science,* 33 (August, 1967), pp. 374-388.

Hahn, Russell H. "Voluntary Standardization and ASAE." *Agricultural Engineering,* April, 1970, pp. 231-232.

Hodges, L.H. "The Voluntary Standards Program for Agricultural Tractors." U.S Department of Transportation, *Agricultural Tractor Safety on Public Roads and Farms.* Washington: U.S. Government Printing Office, 1971, pp. A-123-138.

"How Deere Makes Hay With Europe's Farmers," *Business Abroad,* September 5, 1966, pp. 10, 38-39.

Jetter, Karl. "Die Landmaschinenindustrie Konzentriert." *Frankfurter Allegemeine Zeitung,* May 21, 1970.

Kelleher, Grant W. "The Common Market Antitrust Laws: The First Ten Years," *Antitrust Bulletin,* 12 (Winter, 1967), pp. 1219-1252.

Kudrle, Robert T., "A 'Reverse Dumping Duty' for Canada?" *Canadian Journal of Economics,* 7 (February, 1974), pp. 75-81.

Lanzillotti, Robert F. "The Automobile Industry," In Walter Adams, ed., *The Structure of American Industry.* 4th ed. New York: The Macmillan Company, 1971.

Nieuwenhuysen, J.P. "The Trade Practices Act: Recent Developments and Some Proposals for Change," *Australian Economic Review,* 4th Quarter, 1969, pp. 19-24.

Phillips, W.G. "The Farm Machinery Industry." In John R. Moore and

Richard G. Walsh, eds., *Market Structure of the Agricultural Industries.* Ames: The Iowa State University Press, 1966, pp. 324–357.

Prest, A.R., and Turvey, Ralph. "Cost-Benefit Analysis, A Survey." *Economic Journal,* 75 (December, 1965), pp. 685–735.

Rayner, A. J., and Cowling, K. "Demand for a Durable Input: An Analysis of the United Kingdom Market for Farm Tractors." *Review of Economics and Statistics,* 49 (November, 1967), pp. 590–598.

"Riding the Farm Boom." *Business Week,* October 27, 1973, pp. 74–78.

Skromme, Arnold B. "The Growth of ASAE and the Farm Equipment Industry." *Agricultural Engineering,* April, 1970, pp. 181–184.

Tanquary, E.W. "Standardization: World-Wide." *Agricultural Engineering,* September, 1963, pp. 486–487, 496.

"UAW Sees Peril in Foreign Parts." *Business Week,* July 11, 1970, p. 80.

"U.S. Tractors Plow Some Global Fields." *Business Week,* October 24, 1964, pp. 174–176.

Vernon, Raymond. "International Investment and International Trade in the Product Cycle." *Quarterly Journal of Economics,* 80 (May, 1966), pp. 190–206.

Wenders, John T. "Entry and Monopoly Pricing." *Journal of Political Economy,* 75 (October, 1967), pp. 755–760.

Williamson, Oliver E. "Economies as an Antitrust Defense." *American Economic Review,* 58 (March, 1968), pp. 18–36; "Correction and Reply." *American Economic Review,* 58 (December, 1968), pp. 1372–1376; "Economies as an Antitrust Defense: Reply." *American Economic Review,* 59 (December, 1969), pp. 954–969.

"The World's Top 200 Corporations." *Successo.* December, 1968, pp. 116–117.

PERIODICALS

The following periodicals were used extensively in the preparation of the study and individual articles are not cited:

Agricultural Machinery Journal
L'Argus de l'Automobile
Farm and Power Equipment
Farm Engineering Industry
Farm Implement and Machinery Review
Farm Mechanization
L'Informatore Agrario
Implement and Tractor
Landmaschinen Markt
Power Farming (U.K.)
Producer's Review
Schlepper-Dienst
Tracteurs et Machinisme Agricole

GOVERNMENT PUBLICATIONS

Australia. Commonwealth Bureau of Census and Statistics. *Sales of New Tractors, December Quarter 1970.* Canberra: Commonwealth Bureau of Census and Statistics, 1970.

Australia. Commonwealth Bureau of Census and Statistics. *Statistical Bulletin, Tractors on Rural Holdings.* Canberra: Commonwealth Bureau of Census and Statistics, various annual issues.

Australia. *Manufacturing Industry.* Canberra: Central Office of the Commonwealth Bureau of Census and Statistics, various annual issues.

Australia. *Rural Industries.* Canberra: Central Office of the Commonwealth Bureau of Census and Statistics, various annual issues.

Canada. House of Commons Standing Committee on Agriculture and Colonization. *Minutes of Proceedings and Evidence.* Nos. 1–13, 24th Parliament, 3rd session, 1960.

Canada. Royal Commission on Farm Machinery. *Demand for Farm Machinery–Western Europe,* by Henry G. Scott and David J. Smyth. Royal Commission on Farm Machinery Study No. 9. Ottawa: Queen's Printer, 1970.

Canada. Royal Commission on Farm Machinery. *Farm Machinery Capacity: An Economic Assessment of Farm Machinery Capacity in Field Operations,* by Graham F. Donaldson. Royal Commission on Farm Machinery Study No. 10. Ottawa: Queen's Printer, 1970.

Canada. Royal Commission on Farm Machinery. *Farm Machinery Safety: Physical Welfare Effects of the Man-Machine Interaction on Farms,* by Graham F. Donaldson. Royal Commission on Farm Machinery Study No. 1. Ottawa: Queen's Printer, 1969.

Canada. Royal Commission on Farm Machinery. *Farm Machinery Testing: Scope and Purpose in the Measurement and Evaluation of Farm Machinery,* by Graham F. Donaldson. Royal Commission on Farm Machinery Study No. 8. Ottawa: Queen's Printer, 1970.

Canada. Royal Commission on Farm Machinery. *Farm Tractor Production Costs: A Study in Economies of Scale,* by Neil B. MacDonald, William F. Barnicke, Francis W. Judge, and Karl E. Hansen. Royal Commission on Farm Machinery Study No. 2. Ottawa: Queen's Printer, 1969.

Canada. Royal Commission on Farm Machinery. *Farmers' Attitudes to Farm Machinery Purchases: A Survey Conducted in the Prairie Provinces in Mid-1967,* by Alexander Segall. Royal Commission on Farm Machinery Study No. 4. Ottawa: Queen's Printer, 1969.

Canada. Royal Commission on Farm Machinery. *Locational Advantages in the Farm Machinery Industry: A Comparative Analysis Between Selected Locations in Canada and the United States,* by Neil B. MacDonald. Royal Commission on Farm Machinery Study No. 6. Ottawa: Queen's Printer, 1970.

Canada. Royal Commission on Farm Machinery. *Oligopoly in the Farm Machinery Industry,* by David Schwartzman. Royal Commission on Farm Machinery Study No. 12. Ottawa: Information Canada, 1970.

Canada. Royal Commission on Farm Machinery. *The Prairie Farm Machinery Co-operative,* by Rubin Simkin. Royal Commission on Farm Machinery Study No. 5. Ottawa: Queen's Printer, 1970.

Canada. Royal Commission on Farm Machinery. *Productivity in the Farm Machinery Industry: A Comparative Analysis Between the United States and Canada,* by Christopher J. Maule, Royal Commission on Farm Machinery Study No. 3. Ottawa: Queen's Printer, 1969.

Canada. Royal Commission on Farm Machinery. *Report of the Royal Commission on Farm Machinery,* by Clarence L. Barber. Ottawa: Information Canada, 1971.

Canada. Royal Commission on Farm Machinery. *Research and Development in the Farm Machinery Industry,* by Alex G. Vicas. Royal Commission on Farm Machinery Study No. 7. Ottawa: Queen's Printer, 1970.

Canada. Royal Commission on Farm Machinery. *Revenues, Costs, and Profits in the Farm Machinery Industry,* by Donald Martinusen and Bernard B. Barry. Royal Commission on Farm Machinery Study No. 11. Ottawa: Information Canada, 1970.

Canada. Royal Commission on Farm Machinery. *Special Report on Prices of Tractors and Combines in Canada and Other Countries,* by Clarence L. Barber. Ottawa: Queen's Printer, 1969.

France. Centre National d'Etudes et d'Experimentation de Machinisme Agricole. *Bulletin d'Information.* Antony: Ministère de l'Agriculture, Direction Générale de l'Espace Rural, various issues.

France. Direction Générale des Douanes et Droits Indirects. *Statistiques du Commerces. Extérieur.* various annual issues.

Germany. Statistiches Bundesamt. *Der Aussenhandel der Bundesrepublik Deutschland,* various annual issues.

Italy. Ministero del Commercio con l'Estero. *Machines Agricoles Productrices d'Energie.* Rome: Istituto Nazionale per I Commercio Estero, 1970.

Italy. Ministero del Commercio con l'Estero. *Statistica Annuale del Commercio con l'Estero,* various annual issues.

Organisation for Economic Co-operation and Development. *Development of Farm Motorisation and Consumption and Prices of Motor Fuels in Member Countries.* Paris: Organisation for Economic Co-operation and Development, 1963.

Organisation for Economic Co-operation and Development. *Standard Code for the Testing of Agricultural Tractors.* Paris: Organisation for Economic Co-operation and Development, 1970.

U.K. *Overseas Trade Accounts of the United Kingdom.* London: Her Majesty's Stationery Office, various annual issues.

United Nations. Economic Commission for Europe. *The European Tractor Industry in the Setting of the World Market.* Industry and Materials Committee, 1952. (Mimeographed.)

United Nations. Food and Agriculture Organization. *Production Yearbook.* Volume 24, 1970.

U.S. Bureau of the Census. *Annual Survey of Manufactures.* Washington: U.S. Government Printing Office, various annual issues.

U.S. Bureau of the Census. *Current Industrial Reports.* Series M35S. Washington: U.S. Government Printing Office.

U.S. Bureau of the Census. *U.S. Imports,* Series FT 135, various monthly issues.

U.S. Congress. House. Committee on Agriculture. *Farm Cost-Price Squeeze, Hearings.* Washington: U.S. Government Printing Office, 1961.

U.S. Department of Agriculture. Economic Research Service. *Demand for Farm Tractors in the United States—A Regression Analysis,* by Austin Fox. Washington: U.S. Government Printing Office, 1966.

U.S. Department of Labor. *Handbook of Labor Statistics.* Washington: U.S. Government Printing Office, various annual issues.

U.S. Department of Transportation. *Agricultural Tractor Safety on Public Roads and Farms.* Washington: U.S. Government Printing Office, 1971.

U.S. Federal Trade Commission. *Report on the Agricultural Implement and Machinery Industry.* Washington: U.S. Government Printing Office, 1938.

U.S. Federal Trade Commission. *Report on the Manufacture and Distribution of Farm Implements.* Washington: U.S. Government Printing Office, 1948.

U.S. Temporary National Economic Committee. *Investigation of Concentration of Economic Power,* Monograph No. 36 (prepared by the United States Federal Trade Commission). Washington: U.S. Government Printing Office, 1940.

Notes

NOTES TO CHAPTER ONE

1. This scheme for industrial study is developed in Joe S. Bain, *Industrial Organization* (2nd ed.; New York: John Wiley & Sons, Inc., 1968). The most concise presentation of the approach is Richard E. Caves, *American Industry: Structure, Conduct, Performance* (hereinafter referred to as *American Industry*) (3rd ed.; Englewood Cliffs, New Jersey: Prentice-Hall, Inc., 1972).

2. During some of the later years of the study, Spain, with a tractor park of about 200,000, was both producing and absorbing tractors in greater numbers than was Australia. The most persuasive reason for omitting Spain from the study is that tractor-making there is a highly protected industry which is, in effect, isolated internationally except for participation by several foreign firms. The most important of these is the Chrysler Corporation which is not a participant in any of the other markets as a tractor-maker but rather is involved in the Spanish diesel engine industry. *Agricultural Machinery Journal* (hereinafter referred to as *AMJ*), June, 1970, p. 45. Tractor manufacturing in Spain really has the character of a "development project" which possibly (although not necessarily) can be defended either on "infant industry" grounds or for other reasons such as positive externalities, and which make it necessary to go well beyond industrial economics for proper evaluation. The same reasoning applies to smaller but also rapidly growing tractorization schemes in Mexico and several South American countries, although neither their absorption nor their park approached the size of the seven countries considered in the present study by 1970. As in the Spanish case, their high costs and low volume might be justified with a broader set of criteria than industrial economics and international trade analysis usually employ.

3. Joan Robinson, *The Economics of Imperfect Competition* (London: Macmillan & Co. Ltd., 1933), p. 17.

4. Bain, *Industrial Organization*, p. 6.

5. The approach to "industry" based on perceived interdependence is developed in William Fellner, *Competition Among the Few* (reprinted; New

York: Augustus Kelley, 1965), pp. 3-44. See also Thomas C. Schelling, *The Strategy of Conflict* (Cambridge: Harvard University Press, 1960), pp. 21-28, 53-80.

6. F.M. Scherer, *Industrial Market Structure and Economic Performance* (hereinafter referred to as *Market Structure*) (Chicago: Rand McNally & Company, 1970), pp. 53-54.

7. *Farm Implement and Machinery Review* (hereinafter referred to as *FIMR*), June 1, 1967, p. 556.

8. Richard E. Caves, "Foreign Investment, Trade, and Industrial Growth" (hereinafter referred to as "Foreign Investment"), Royer Lectures, University of California, Berkeley, December 1-2, 1969, pp. 6, 7. (Mimeographed.)

9. The early estimates of scale economies are from Joe S. Bain, *Barriers to New Competition* (hereinafter referred to as *Barriers*) (Cambridge: Harvard University Press, 1956). The later material is developed in Neil B. MacDonald, William F. Barnicke, Francis W. Judge, and Karl E. Hansen, *Farm Tractor Production Costs: A Study in Economies of Scale* Ottawa: Queen's Printer, 1969).

10. Material in this paragraph is from company annual reports, various editions of Moody's *Industrial Manual* and "The World's Top 200 Corporations," *Successo*, December, 1968, pp. 116-117.

11. *Implement and Tractor* (hereinafter referred to as *I&T*), July 7, 1967, p. 21.

12. David Schwartzman, *Oligopoly in the Farm Machinery Industry* (hereinafter referred to as *Oligopoly*) (Ottawa: Information Canada, 1970), Tables 2-1 and 2-2, pp. 7, 8; *AMJ*, July, 1969, p. 81; Donaldson, Lufkin and Jenrette, Inc., *The European Agricultural Equipment Industry and Competitive Positions of North American Producers* (hereinafter referred to as *European Equipment Industry*) (New York: Donaldson, Lufkin and Jenrette, Inc., 1966), p. 11. (Mimeographed.); M. Bernasek and A.M. Kubinski, "Agricultural Machinery and Implements," in Alex Hunter, ed., *The Economics of Australian Industry* (Melbourne: Melbourne University Press, 1963), p. 461.

13. David Schwartzman, *Oligopoly*, p. 126.

14. Corwin D. Edwards, "Size of Markets, Scale of Firms, and the Character of Competition," in E.A.G. Robinson, ed., *Economic Consequences of the Size of Nations* (London: Macmillan & Co. Ltd., 1960), p. 125.

15. Bain, *Industrial Organization*, pp. 137-144.

16. Caves, "Foreign Investment," pp. 20-22, and "International Corporations: The Industrial Economics of Foreign Investment" (hereinafter referred to as "International Corporations"), *Economica*, 38 (February, 1971), pp. 12-14.

17. Caves's discussion of this advantage stresses the international firm's ability to use internally generated funds, to enjoy a high credit rating, and probably to have access to a broader range of capital markets. It is possible that there could be yet another reason. As Caves has noted elsewhere (*American Industry*, p. 27), "Even a borrower with a much better credit rating, such as a large going firm in another industry, might have to pay more interest than would an existing steel firm planning to expand." To the extent that entry into a new geographic area is regarded by the capital market merely as an extension of activities of which the firm has already proven itself capable, it will enjoy an advantage; against this

must be weighed whatever might be regarded as formidably "foreign" about the new operating environment.

18. Bain, *Industrial Organization*, p. 254.

19. Both sets of hypotheses have been assembled by Caves along with many original suggestions in "Foreign Investment" and "International Corporations."

20. Bain, *Industrial Organization*, pp. 419, 422.

21. Federal Trade Commission, *Report on the Agricultural Implement and Machinery Industry: Concentration and Competitive Methods* (hereinafter referred to as *Agricultural Implement Industry*) (Washington: U.S. Government Printing Office, 1938); Federal Trade Commission, *Report on Manufacture and Distribution of Farm Implements* (hereinafter referred to as *Manufacture and Distribution*) (Washington: U.S. Government Printing Office, 1948).

22. Political and Economic Planning, *Agricultural Machinery* (London: Political and Economic Planning, 1949).

23. United Nations Economic Commission for Europe, Industry and Materials Committee, *The European Tractor Industry in the Setting of the World Market* (hereinafter referred to as *European Tractor Industry*), 1952. (Mimeographed.)

24. Michael Conant, "Aspects of Monopoly and Price Policies in the Farm Machinery Industry Since 1902" (unpublished Ph.D dissertation, University of Chicago, 1949); Warren Wright Shearer, "Competition Through Merger: An Economic Analysis of the Farm Machinery Industry" (unpublished Ph.D dissertation, Harvard University, 1951); Elvis L. Eckles, "The Development of Oligopoly in the Farm Implement Industry" (unpublished Ph.D dissertation, University of Illinois, 1953).

25. Syndicat des Constructeurs Français de Matériels de Motoculture, *L'Industrie des Tracteurs et des Motoculteurs* (Paris: Société Nouvelle Mercure, 1955).

26. W.G. Phillips, *The Agricultural Implement Industry in Canada* (Toronto: University of Toronto Press, 1956).

27. R.E. Linneman, "The United States Tractor Industry in Selected Foreign Markets" (hereinafter referred to as "U.S. Tractor Industry") (unpublished Ph.D. dissertation, University of Illinois, 1964).

28. E.P. Neufeld, *A Global Corporation* (hereinafter referred to as *Global Corporation*) (Toronto: University of Toronto Press, 1969).

29. Simon N. Whitney, *Antitrust Policies: American Experience in Twenty Industries*, Volume II, *Famous Antitrust Cases* (New York: The Twentieth Century Fund, 1958), pp. 227–256; W.G. Phillips, "The Farm Machinery Industry," in John R. Moore and Richard G. Walsh, eds., *Market Structure of the Agricultural Industries* (Ames: The Iowa State University Press, 1966); M. Bernasek and Z.M. Kubinski, "Agricultural Machinery and Implements" (hereinafter referred to as "Agricultural Machinery"), in Alex Hunter, ed., *The Economics of Australian Industry* (Melbourne: Melbourne University Press, 1963).

30. The Royal Commission on Farm Machinery's complete works, some of which will hereinafter be cited by shortened titles (given in parentheses), are as follows: Graham F. Donaldson, *Farm Machinery Safety: Physical Welfare Effects of the Man-Machine Interaction on Farms* (*Farm Machinery Safety*),

RCFM Study No. 1 (Ottawa: Queen's Printer, 1968); Neil B. MacDonald, William F. Barnicke, Francis W. Judge, and Karl E. Hansen, *Farm Tractor Production Costs: A Study in Economies of Scale* (*Tractor Costs*), RCFM Study No. 2 (Ottawa: Queen's Printer, 1969); Christopher J. Maule, *Productivity in the Farm Machinery Industry: A Comparative Analysis Between the United States and Canada* (*Productivity in the Farm Machinery Industry*), RCFM Study No. 3 (Ottawa: Queen's Printer, 1969); Alexander Segall, *Farmers' Attitudes to Farm Machinery Purchases: A Survey Conducted in the Prairie Provinces in Mid-1967* (*Farmers, Attitudes*), RCFM Study No. 4 (Ottawa: Queen's Printer, 1969); Clarence L. Barber, *Special Report on Prices of Tractors and Combines in Canada and Other Countries* (*Prices*), (Ottawa: Queen's Printer, 1969); Rubin Simkin, *The Prairie Farm Machinery Co-operative: The Canadian Co-operative Implements Limited* (*Prairie Co-operative*), RCFM Study No. 5 (Ottawa: Queen's Printer, 1970); Neil B. McDonald, *Locational Advantages in the Farm Machinery Industry: A Comparative Analysis Between Selected Locations in Canada and the United States* (*Locational Advantages*), RCFM Study No. 6 (Ottawa: Queen's Printer, 1970); Alex G. Vicas, *Research and Development in the Farm Machinery Industry* (*Research and Development*), RCFM Study No. 7 (Ottawa: Queen's Printer, 1970); Graham F. Donaldson, *Farm Machinery Testing: Scope and Purpose in the Measurement and Evaluation of Farm Machinery*, RCFM Study No. 8 (Ottawa: Queen's Printer, 1970); Henry G. Scott and David J. Smyth, *Demand for Farm Machinery—Western Europe* (*Demand for Farm Machinery*), RCFM Study No. 9 (Ottawa: Queen's Printer, 1970); Graham F. Donaldson, *Farm Machinery Capacity: An Economic Assessment of Farm Machinery Capacity in Field Operations*, RCFM Study No. 10 (Ottawa: Queen's Printer, 1970); Donald Martinusen and Bernard P. Barry, *Revenues, Costs, and Profits in the Farm Machinery Industry* (*Revenues, Costs, and Profits*), RCFM Study No. 11 (Ottawa: Information Canada, 1970); David Schwartzman, *Oligopoly in the Farm Machinery Industry* (*Oligopoly*), RCFM Study No. 12 (Ottawa: Information Canada, 1970); Clarence L. Barber, *Report of the Royal Commission on Farm Machinery* (*Report*), (Ottawa: Information Canada, 1971).

31. See Lawrence J. White, *The Automobile Industry Since 1945* (hereinafter referred to as *Automobile Industry*) (Cambridge: Harvard University Press, 1971), pp. 3–4; Scherer, *Market Structure*, pp. 5–6. The point is conceded in Bain, *Industrial Organization* (pp. 364–365), though not emphasized as much as in many other writings.

NOTES TO CHAPTER TWO

1. See, for example, Caves, *American Industry*, pp. 30–31.
2. Both possibilities are examined by Bain, *Barriers*, pp. 105–106, but he appears more strongly inclined to make the "pessimistic" assumption. This is also the view of the co-originator of the economies of scale barrier, Paolo Sylos-Labini. Their views are presented with some additional insights in Franco Modigliani, "New Developments on the Oligopoly Front," *Journal of Political*

Economy, 66 (June 1958), pp. 215–232. For a critique, see Scherer, *Market Structure*, p. 228.

3. *I&T*, July 28, 1956, pp. 32–33, 69; March 7, 1965, p. 22; March 21, 1965, p. 23; Australia, Commonwealth Bureau of Census and Statistics, *Sales of New Tractors, December Quarter 1970* (Canberra: Commonwealth Bureau of Census and Statistics, 1970), p. 7.

4. United Nations, *European Tractor Industry*, pp. 5, 6; Linneman, "United States Tractor Industry," p. 100.

5. United Nations, *European Tractor Industry*, pp. 5, 6.

6. Zvi Griliches, "The Demand for a Durable Input: Farm Tractors in the United States, 1921–57," in A.C. Harberger, ed., *The Demand for Durable Goods* (Chicago: The University of Chicago Press, 1960), pp. 181–207. W.A. Cromarty, *The Demand for Farm Machinery and Tractors* (East Lansing: Michigan State University Press, 1959). A.J. Rayner, *An Econometric Analysis of the Demand for Farm Tractors*, University of Manchester, Department of Agricultural Economics, Bulletin No. 113, October, 1966. (Mimeographed.) A.J. Rayner and K. Cowling, "Demand for a Durable Input: An Analysis of the United Kingdom Market for Farm Tractors," *Review of Economics and Statistics*, 44 (November, 1967), pp. 590–598. Austin Fox, *Demand for Farm Tractors in the United States—A Regression Analysis*, Economic Research Service, U.S. Department of Agriculture, Agricultural Economic Report, No. 103 (Washington: U.S. Government Printing Office, 1966); Scott and Smyth, *Demand for Farm Machinery*.

7. Scott and Smyth, *Demand for Farm Machinery*, pp. 54–55.

8. Earl O. Heady and Luther C. Tweeten, *Resource Demand and Structure of the Agricultural Industry* (Ames: Iowa State University Press, 1963), p. 321; Cromarty, *The Demand for Farm Machinery and Tractors*.

9. Compare Table 2-1 with Table A–6 in Schwartzman, *Oligopoly*, p. 223.

10. Material in this paragraph from the following sources: Farm Equipment Institute, *Facts, 1959 Edition* (Chicago: The Farm Equipment Institute, 1959), p. 7; *I&T*, February 21, 1972, p. 30; *AMJ*, October, 1969, p. 31; Centre National d'Etudes et d'Experimentation de Machinisme Agricole, *Bulletin d'Information*, (Antony: Ministère de l'Agriculture, Direction Générale de l'Espace Rural), June-July 1971, p. 55; *Landmaschinen Markt*, October 1, 1969, p. 55; *AMJ*, May, 1969, p. 49; Commonwealth Bureau of Census and Statistics, *Sales of New Tractors, December Quarter 1970* (Canberra: Commonwealth Bureau of Census and Statistics, 1970), p. 6.

11. Scott and Smyth, *Demand for Farm Machinery*, p. 130; FAO *Production Yearbook*, Volume 24, 1970, pp. 10, 11.

12. Barber, *Report*, pp. 473–474.

13. *I&T*, February 21, 1972, p. 30; liquified propane (L.P.) enjoyed a brief burst of popularity in the U.S. in the fifties, peaking at 5.5 percent of the unit market in 1957, but it had declined to less than 1 percent in 1970; although the fuel was clean-burning and allowed farmers in some areas to use the same fuel for their tractors and home-heating, machines converted to L.P. shared with diesels the hard-starting problem in cold weather, and the relative price of L.P. in

many areas became less favorable over time. E.M. Dieffenbach and R.B. Gray, "The Development of the Tractor," in United States Department of Agriculture, *Yearbook of Agriculture, 1960* (Washington: U.S. Government Printing Office, 1960), p. 37, and Wayne H. Worthington, *50 Years of Tractor Development* (New York: Society of Automotive Engineers, 1966), p. 14.

14. Commonwealth Bureau of Census and Statistics, *Statistical Bulletin, Tractors on Rural Holdings* (Canberra: Commonwealth Bureau of Census and Statistics, various issues).

15. C.N.E.E.M.A., *Bulletin d'Information,* various issues.

16. Organization for Economic Co-operation and Development, *Development of Farm Motorisation and Consumption and Prices of Motor Fuels in Member Countries* (Paris: Organization for Economic Co-operation and Development, 1963), p. 66.

17. Linneman, "U.S. Tractor Industry," pp. 39–41.

18. European gasoline engines availability was due to their use in powering cars and at least some trucks. The gasoline models produced for Massey-Ferguson by Standard Motors in the U.K. in the late fifties sold domestically for about the same percentage discount from their diesel counterparts as those sold in North America (although the British prices were more than 25 percent lower for both models).

19. *I&T,* May 19, 1956, p. 60; October 21, 1963, p. 28.

20. Barber, *Report,* pp. 192–194

21. A.J. Rayner, *An Econometric Analysis of the Demand for Farm Tractors,* p. 5; *AMJ,* December, 1969, p. 33.

22. Warren Baum, *The French Economy and the State,* Rand Corporation Research Study (Princeton: Princeton University Press, 1958), pp. 297–300; *Terre Moderne* (Paris: Les Tracteurs Agricoles Renault, 1950), pp. 354–55; *FIMR,* May, 1, 1957, p. 101; April 1, 1958, p. 1855; February 1, 1959, p. 1489; May 1, 1959, p. 107; *Landmaschinen Markt,* August 2, 1967, p. 4; *Farm Mechanization,* April, 1955, p. 36; *FIMR,* March 1, 1961, p. 333.

23. Bernasek and Kubinski, "Agricultural Machinery," p. 478; *Producer's Review,* February, 1965, p. 73; *Producer's Review,* October, 1963, p. 37.

NOTES TO CHAPTER THREE

1. White, *Automobile Industry,* p. 19.

2. Much of the ensuing discussion summarizes the presentation in *Tractor Costs* itself and the discussion of its findings as they appear in Schwartzman, *Oligopoly* pp. 39–64, or in Barber, *Report,* pp. 85–94.

3. In fact, the Commission did not perform its declared exercise properly, in that the estimation task was simplified by using the outside purchase cost for the 90 h.p. tractor plus the weighted average in-house production costs of the three different sized machines. In order to keep our estimates comparable with the Commission's at this point, the error is retained in all of the estimates presented in Table 3–1. An attempt to correct for the overestimation (amounting to $84 per unit at 60,000 units) is, however, reflected in Table 3–4, which compares U.S. and British cost estimates.

4. The Commission study seems undecided about the opportunity cost of the funds used for the project. An advance discussion of the procedure of the *Tractor Costs* states:

> When considering the purchase of additional productive equipment (beyond the equipment in general industry use at 60,000 units in North America), the analysts used as a requirement for its justification a pre-tax rate of return of 20 per cent. This figure is more commonly used to evaluate individual investment alternatives than is the minimum cost of capital (7.5 per cent). The basis is that marginal investments should produce a return after taxes that is substantially above the market costs of the funds involved. The additional amount may be considered to cover the risks and uncertainties in the decision and to perform a rationing function for capital projects (MacDonald et al, *Tractor Costs*, p. 21.)

That the minimum cost of capital to the firm is a function of the amount borrowed (or borrowed and raised through equity issues) and rises with that amount, in part because management has subjective notions of cost related to risk which may manifest themselves in an after-tax target of something like 10 percent for a marginal project, would seem to suggest that the opportunity cost to the firm of the funds involved in new projects should therefore be 10 percent rather than the 7.5 percent used for the cost of capital for the bulk of plant expenditures.

5. MacDonald et al., *Tractor Costs*, p. 146.

6. Barber, *Prices*, p. 68.

7. Of interest in this connection is that the automatic handling of engine parts between machining stations, which White finds appropriate for volumes no lower than 260,000 units per year in the automobile industry, was actually found to be justified by *Tractor Costs* for all three tractor volumes. Though the degree of mechanization may be lower than with the "automatic handling" described by White, neither illustrations nor descriptions of the line leave this impression.

> Special purpose machining lines were specified for such large, complex components as the cylinder block, crankshaft, cylinder head and transmission case. These lines provide for mechanized transfer of components through sequential machining stations . . . Such lines were economically justified for these major components even at the lowest level of 20,000 units (MacDonald et al, *Tractor Costs*, p. 84).

Machining costs do decline 18 percent between 20,000 and 90,000 units, and this contrasts with White's contention that "Once a volume has been attained that warrants automatic handling of the pieces, a fairly proportional relationship between speed and investment in machine tools takes place" (White, *Automobile Industry*, p. 24). It does not contradict his conclusion, however, that the proportional relation begins at an output in excess of 250,000 units per annum. Both Ford and Massey-Ferguson tractor engines are manufactured (with indusal engines) at this volume or greater, and this fact may go far to illuminate the otherwise unexplained Commission finding based on industry sources that, by contrast with tractor costs at a 90,000 unit output, ". . . we have estimated that an additional saving of $150 (U.S.) could be obtained on the higher volume production; particularly of diesel engines available to Ford and Massey-Ferguson in England" (Barber, *Prices*, p. 68).

8. Schwartzman, *Oligopoly*, pp. 60–63.

9. Schwartzman, *Oligopoly*, p. 63. This conclusion appears to be contradicted by a procedure used in *Prices*, pp. 199–200, where outside purchased parts estimates are adjusted downward by 20 percent. Not only is this procedure unmotivated but Schwartzman and Barber's *Report* simply ignore it.

10. Schwartzman concludes that the former explanation is unpersuasive because the Fabricated Metals group from which most of the purchasing is done is a subsector of only modest profitability. While almost any measure of returns indicates that this is true, the subsector is enormously heterogeneous, and there could be substantial profits on some of the products sold to tractor manufacturers. Schwartzman is left with the explanation that vendor inefficiency must be the culprit. Part of the answer may lie here because, while the age of equipment in these activities is no older than average over the entire economy, the comparison is with in-house fabrication from a totally new plant. Further, as with monopoly considerations, aggregate figures may disguise considerable variation.

11. Bain, *Barriers*, p. 249.

12. Bain defines a "relatively flat" curve as one which "will find unit cost only 1 or 2 per cent higher than optimal at half optimal scales; a moderately sloped curve will find them 4 or 5 per cent higher at half optimal scales; a relatively steep curve will find them 8 to 10 per cent higher" (*Ibid.*, p. 161). With respect to tractors his findings indicate "costs only slightly higher at half the minimum optimal scale and only moderately higher at a fifth or a tenth of it" (*Ibid.*, p. 249).

13. The relative prices of labor and capital goods also affect the comparison, of course. For a discussion of this issue, see section on "Progressiveness in Manufacture" in Chapter Twelve.

14. For a definition of X-inefficiency and by implication that of X-efficiency as well, see p. 274.

15. Detailed production material by country is presented in Chapter Eleven.

16. MacDonald et al., *Tractor Costs*, p. 161.

17. *Ibid.*, p. 284–285.

18. This inference is also drawn by the Commission staff, although there is no attempt to estimate savings from concentrating on fewer than three models differing widely in size. See also Barber, *Prices*, p. 68.

19. MacDonald et al., *Tractor Costs*, p. 10.

20. Barber, *Report*, p. 101.

21. MacDonald et al., *Tractor Costs*, pp. 162–165.

22. For a discussion of the contribution of concentration and the ratio of fixed to variable costs to the likelihood of "cut-throat competition," see R.E. Caves, *Air Transport and Its Regulators* (hereinafter referred to as *Air Transport*) (Cambridge: Harvard University Press, 1962), p. 79.

23. Barber, *Prices*, pp. 182–200.

24. *Ibid.*, p. 68. Dunning's pioneering study of American investment in Britain gives questionnaire evidence of two agricultural machinery firms manufacturing mainly tractors which as early as 1953–54 experienced "little difference (in labor productivity) though favoring U.S. company." British output was also cheaper overall. John H. Dunning, *American Investment in British Manufacturing Industry* (London: George Allen & Unwin Ltd, 1958), pp. 322, 343, 344.

25. *Ibid.*, p. 199.

26. *I&T*, January 2, 1954, p. 85.

27. Neufeld, *Global Corporation*, p. 175.

28. *I&T*, May 28, 1960, p. 30.

29. U.S. Congress, House, Committee on Agriculture, *Farm Cost-Price Squeeze, Hearings*, before the Subcommittee on Equipment, Supplies, and Manpower of the Committee on Agriculture, House of Representatives, 87th Congress, 1st session, 1961, p. 435.

30. *Ibid.*, pp. 544–48.

31. Linneman, "U.S. Tractor Industry," p. 127.

32. *Ibid.*, p. 52.

33. *Ibid.*, p. 122.

34. MacDonald, *Locational Advantages*, p. 27, and Barber, *Report*, p. 308.

35. MacDonald, *Locational Advantages*, p. 124.

36. White, *Automobile Industry*, p. 24.

37. The gasoline-powered Volvo tractor of the late fifties used a "slightly modified" Volvo auto engine, *I&T*, October 5, 1957, p. 83. British Leyland's small tractor in the late sixties used a "B" type diesel engine of the kind used on its small commercial vehicles. *AMJ*, February, 1969, p. 47.

38. This figure was obtained by comparing the Royal Commission on Farm Machinery's estimates as presented in Barber, *Prices*, pp. 4, 12.

39. Schwartzman, *Oligopoly*, pp. 58, 71.

40. *Ibid.*, p. 67.

41. *Ibid.*, pp. 67–70.

42. The 8 percent discount rate is used for tractor costs.

43. Barber, *Report*, pp. 99–100.

44. See pp. 133–136.

45. These figures are drawn from the appendices of MacDonald, et al., *Tractor Costs*.

46. Bain, *Barriers*, p. 158.

47. These figures are based on the "average" tractor transfer price of $4000 which corresponds to a wholesale price of $5,058 according to Commission estimates of industry practice. Five percent of this figure is estimated industry inventory holding cost for large firms (from Table 4-1); at an interest rate of 7.5 percent, this implies invested capital of $3,372 per tractor (all figures in U.S. dollars).

48. This is a conservative estimate which fails to take fully into account the extent to which distribution assets may have to be relatively larger for small firms selling to small dealers with low turnover.

49. Martinusen and Barry, *Revenues, Costs, and Profits*, p. 110.

50. There is a sense in which certain distribution advantages seem as appropriately regarded as belonging to "absolute costs" as to "product differentiation." See p. 276.

NOTES TO CHAPTER FOUR

1. Except where specifically noted, this section is drawn from Dieffenbach and Gray, "Development of the Tractor," and Worthington, *50 Years.*

2. It was impossible at this time for the tractor to be called simply "Ford" because a forward-looking firm in Minneapolis had already appropriated the name for its machine. This firm was defunct by the time of Ford's second entry into the U.S. market in 1939.

3. Dieffenbach and Gray, "Development of the Tractor," p. 34.

4. Worthington, *50 Years,* p. 6.

5. *Producer's Review,* June, 1957, p. 96.

6. *Ibid.*

7. Neufeld, *Global Corporation,* pp. 96–98.

8. Worthington, *50 Years,* p. 13.

9. A thorough discussion of transmission developments can be found in Bruce D. Narsted, "Tractor Technical Development," in Vicas, *Research and Development,* Appendix B, pp. 85–88.

10. *Ibid.,* pp. 88–90.

11. See Chapter Nine.

12. It should also be noted that, for reasons which are explained in Chapter Ten, there has not usually been an important problem of compatibility between tractors of one make and implements produced by others. The interface has either been simple or it has been subject to voluntary industry standardization.

13. Here the leading German manufacturer, Deutz, provided a minor exception with its air-cooled diesels.

14. A minor force working in the opposite direction from international similarity has been the development of different safety regulations for tractors used in different countries, such as maximum allowable speeds and different lighting requirements. These national differences are cited by Baldwin as an example of regulations which hinder international trade, but their actual importance in rendering tractor trade more difficult or expensive is very slight. Robert E. Baldwin, *Non-Tariff Distortions of International Trade* (Washington, D.C.: The Brookings Institution, 1970), pp. 144–145.

15. Most patents have been rather freely cross-licensed in any event. See p. 145.

16. Worthington, *50 Years,* p. 11.

17. *Ibid.,* p. 22.

18. White, *Automobile Industry,* p. 211.

19. Caves, *Air Transport,* p. 49. Caves coins the term in an attempt to avoid ambiguity in the use of "product differentiation." The distinction is between a "pure" product characteristic and a barrier to entry. The latter can be at the same time an important element of structure and the fruit of sustained conduct.

20. R.B. Gray, *Development of the Agricultural Tractor in the United States,*

Part I (St. Joseph, Michigan: American Society of Agricultural Engineers, 1956), pp. 31-34.

21. The procedures and measurements in the early seventies can be found in the *Implement and Tractor Redbook,* January 31, 1972, pp. A125-A152.

22. OECD *Standard Code for the Testing of Agricultural Tractors* (Paris: Organization for Economic Co-operation and Development, 1970).

23. The operational notion Bain attached to his categories related to the price disadvantage of a newcomer selling a similar product. This disadvantage was not specifically estimated for tractors but the overall "great" barrier category implies at least a "5 percent of price disadvantage for 10 years or a 10 percent of price disadvantage for five years," for a firm entering at the scale of one optimum plant. For the tractor industry this implies such a huge share of any or all markets that the disadvantage must be far greater. Bain, *Barriers,* pp. 126-127.

24. *Ibid.,* p. 129.

NOTES TO CHAPTER FIVE

1. Material used to compare automobile industry distribution practices with those in farm machinery are drawn from White, *Automobile Industry,* pp. 136-155.

2. Martinusen and Barry, *Revenues, Costs, and Profits,* p. 103.

3. White, *Automobile Industry,* pp. 152-155.

4. Dun and Bradstreet, Inc., *Key Business Ratios* (New York: Dun and Bradstreet, Inc.).

5. See White, *Automobile Industry,* pp. 154 and 328 for a discussion of this measure of risk.

6. $\beta = .66$, $R^2 = .77$, $t = 6.58$.

7. White's results are in fact stronger than he states them because a slip in calculation causes him to show the Standard Deviation/Average ratio for profits (without officers' salaries) to be twice that for the auto companies when the true figure is approximately three times. White, *Automobile Industry,* p. 153.

8. Schwartzman, *Oligopoly,* p. 177.

9. This has been a particular problem for the employee-managed Canadian Co-operative Implements, Limited. Simkin, *Prairie Co-operative,* p. 58.

10. Harvard Business School, *Deere and Company,* ICH 13G124R BP 865R, 1966, p. 19.

11. *I&T,* August 7, 1967, p. 63.

12. *I&T,* April 2, 1960, p. 13; May 7, 1963, p. 21.

13. Federal Trade Commission, *Implement and Machinery Industry,* p. 208.

14. *I&T,* December 21, 1970, pp. 12-13; February 7, 1971, p. 14.

15. *I&T,* December 21, 1966, pp. 24-25.

16. *I&T,* January 21, 1964, p. 1; Martinusen and Barry, *Revenues, Costs, and Profits,* p. 121.

17. Farm and Power Equipment, *Why Do Farmers Buy Their Machines Where They Do?* (St. Louis: National Farm and Power Equipment Dealers Association, 1965); *I&T,* April 7, 1970, pp. 20-21; December 1, 1956, pp. 30-31, 59; December 15-29, 1956, pp. 24-26; January 26, 1957, pp. 42-45; Martinusen

and Barry, *Revenues, Costs, and Profits,* p. 121; *Fortune,* July, 1961, p. 188.

18. For unevenly represented firms, this is clearly less of a problem. See p. 116.

19. Barber, *Report,* p. 166.

20. *Ibid.,* p. 174.

21. *I&T,* June 21, 1970, pp. 22–23.

22. White, *Automobile Industry,* pp. 142–44. The possible relevance of the "forcing" model is also suggested by B.P. Pashigian, *The Distribution of Automobiles, An Economic Analysis of the Franchise System* (Englewood Cliffs, New Jersey: Prentice-Hall, Inc., 1961) pp. 33–34, 52–56.

23. Barber, *Report,* p. 165.

24. Harvard Business School, *Deere and Company,* p. 5.

25. *Ibid.,* p. 15.

26. *I&T,* October 31, 1959, p. 11. A development beginning about 1970 was the sale of small Japanese machines which did not compete directly with the range of the major franchisor. See discussion in Chapter Seven.

27. The situation in Saskatchewan in 1967 is given in Barber, *Report,* p. 164; similar patterns are thought to prevail throughout the rest of North America.

28. The legal challenges are well summarized in A.E. Kahn and Joel Dirlam, *Fair Competition* (Ithaca: Cornell University Press, 1954), pp. 110–114; 178–179. In *U.S. v. J.I. Case* in 1948, a small farm equipment company's demonstration that over 70 percent of its dealerships sold competing equipment, not necessarily tractors or combines, and that (in boom post-war conditions) other manufacturers successfully found outlets for their products, won a judgment of its innocence. Had Deere or International Harvester been the defendant the former argument would have looked less persuasive, although the criterion is a very weak one. Material against these firms was prepared, but the Justice Department lost heart when the J.I. Case litigation brought only lukewarm response from the court. The second argument would have been weaker from the mid-fifties onward.

29. The major North American exception has been Ford, which during the sixties greatly increased its implement offerings. See p. 116.

30. Again, this ignores small non-competing tractors.

31. *I&T,* March 31, 1965, pp. 22–23, 64–65.

32. *Fortune,* July, 1961, p. 132.

33. Phillips, *The Agricultural Implement Industry in Canada,* p. 145.

34. *Ibid.,* p. 152. Deere still sold on consignment in Canada in 1970, and most firms there didn't revise the consignment system until 1945, but dealer sales were closely supervised.

35. Actual terms were complicated and differed considerably from company to company. See Federal Trade Commission, *Implement and Machinery Industry,* pp. 291–297.

36. Phillips, *The Agricultural Implement Industry in Canada,* p. 152.

37. For a discussion of the competitive aspects of these practices see pp. 115–116.

38. Barber, *Report,* pp. 193–195, gives the total figure for all machinery sold on time as between 30 and 50 percent. Tractors and combines were undoubtedly

purchased on time far out of proportion to their contribution to final sales.

39. Barber, *Report,* p. 190.

40. See pp. 210–212.

41. This alternative will be discussed in Chapter Seven.

NOTES TO CHAPTER SIX

1. Bain, *Barriers,* p. 170.

2. *Ibid.,* p. 176.

3. See the discussion in Chapter Eight.

4. National Retail Farm Equipment Dealers Association, *Official Guide* (St. Louis: NRFEA Publications, Inc., 1961), p. 98.

5. Barber, *Report,* p. 276.

6. *Ibid.,* p. 275.

7. *Ibid.,* p. 276.

8. Bain, *Industrial Organization,* p. 259.

9. This is not necessarily to argue that entry is "blockaded," i.e., that the full short-run monopoly price can be charged without attracting entry. For this condition to hold, very restrictive assumptions about entry barriers and demand elasticities must hold. See John T. Wenders, "Entry and Monopoly Pricing," *Journal of Political Economy,* 75 (October, 1967), pp. 755–760.

10. This is a point noted in Herbert G. Grubel, "Intra-Industry Specialization and the Pattern of Trade," *Canadian Journal of Economics and Political Science,* 33 (August, 1967), p. 384.

11. Linneman, "U.S. Tractor Industry," p. 52.

12. *Ibid.,* p. 50.

NOTES TO CHAPTER SEVEN

1. Caves, "International Corporations," p. 4.

2. The "export stake" explanation has been put forward by many writers. See *Ibid.,* p. 4.

3. J.N. Behrman, *Some Patterns in the Rise of the Multinational Enterprise* (Chapel Hill: Graduate School of Business, University of North Carolina, 1969), p. 3.

4. Barber, *Report,* p. 128.

5. Michael Conant, "Competition in the Farm Machinery Industry," *Journal of Business,* 26 (January, 1953), p. 36.

6. Neufeld, *Global Corporation,* pp. 103–110.

7. *Ibid.,* pp. 122–125.

8. A full discussion of Ford's behavior is presented in Chapter Eight.

9. *I&T,* November 20, 1947, p. 116.

10. R.B. Gray, *Development of the Agricultural Tractor in the United States, Part II* (St. Joseph, Michigan: American Society of Agricultural Engineers, 1956), pp. 44–52.

11. Phillips, "The Farm Machinery Industry," p. 348.

12. Cockshutt's initial line featured a continuous-running power take-off for

external machinery which allowed the latter to operate independently of power transmitted to the tractor's wheels. This important but rather simple advance was quickly and widely imitated, and it well illustrates the kind of minor innovation which has characterized the development of tractor design. Narsted, "Tractor Technical Development," p. 90.

13. Barber, *Report,* pp. 46–47.

14. *I&T,* May 22, 1954, p. 58.

15. Barber, *Report,* p. 52.

16. *I&T,* March 7, 1970, p. 14.

17. *I&T,* January 9, 1960, p. 70; May 14, 1960, pp. 78–79; May 28, 1960, p. 42; July 23, 1960, p. 59; February 19, 1962, p. 111.

18. In 1960, a 40 h.p. Porsche sold for $3,610 (retail), while the comparable Oliver "550" diesel, was $3,176. For one dealer's experience in selling Porche tractors, see *I&T,* May 28, 1960, p. 28.

19. In 1972 models up to 125 h.p. were added to the line.

20. In 1968 Deutz established majority control of Fahr, the second largest combine producer in Germany and entered into a joint marketing arrangement with it in Europe (see Chapter Eight). To become a full-line firm in the North American market, however, would require developing scores of specialized smaller machines, in addition to a combine.

21. *I&T,* December 21, 1964, pp. 14–15; *AMJ,* January 1969, p. 49.

22. *AMJ,* April, 1971, p. 23.

23. *I&T,* October 19, 1957, pp. 27–28.

24. *I&T,* March, 1960, p. 30.

25. *I&T,* July 1, 1964, p. 24.

26. *I&T,* August 7, 1972, p. 47.

27. Mitsubishi $2,640, Satoh $2,400, Kubota $2,000–$2,500, I–H $3,011. *Fortune,* February, 1971, p. 45.

28. Barber, *Prices,* p. 5.

29. *I&T,* January 2, 1954, pp. 84–85.

30. P.E.P., *Agricultural Machinery,* p. 11.

31. *Ibid.,* pp. 11–17.

32. *Ibid.,* p. 6.

33. *Ibid.,* pp. 16–17; *British Farm Mechanisation,* November, 1950, p. 352.

34. *FIMR,* July 1, 1951, p. 470; July 1, 1952, p. 484.

35. A shortage of steel in 1951 and 1952 had some effect on industry production in general, but it was not seriously damaging. *FIMR,* November 1, 1951, p. 109; March 1, 1952, p. 1751.

36. *FIMR,* December 1, 1960, p. 1209.

37. *FIMR,* October 1, 1967, pp. 932–933.

38. *Farm Mechanization,* October, 1962, p. 344.

39. Perhaps the most unique technique ever used in the merchandising of tractors was tried by Dutra in 1971. The firm guaranteed a trade-in schedule based on age of tractor and number of hours worked (approximately two-thirds of the purchase price guaranteed for an average-worked machine after one year). *AMJ,* July, 1971, p. 64.

40. *AMJ*, April, 1969, p. 68.
41. United Nations, *European Tractor Industry*, pp. 35–36; Syndicat, *L'Industrie des Tracteurs*, pp. 10–15.
42. Linneman, "U.S. Tractor Industry," p. 43.
43. Neufeld, *Global Corporation*, p. 166.
44. Moody's *Industrial Manual*, 1970 edition, pp. 1452, 3308. The name of the firm was changed to Fiat-France and Umberto Agnelli, brother of the Fiat chairman, became the chairman of the board of directors. Beerman's *Financial Yearbook of Europe*, 3rd ed., p. 172.
45. C.N.E.E.M.A., *Bulletin d'Information*, March-April, 1959, p. 9.
46. See Figure 11-2.
47. *FIMR*, March 1, 1960, p. 1624.
48. Harvard Business School, *Deere and Company*, p. 9.
49. Donaldson, et al., *European Implement Industry*, p. 8.
50. C.N.E.E.M.A., *Bulletin d'Information*, March-April, 1968, Annex, p. V; March-April, 1970, Annex, p. V.
51. Price material provided to the author by C.N.E.E.M.A.
52. *Ibid.*
53. Allis-Chalmers, *Annual Report*, 1960, p. 1.
54. *Ibid.*
55. Allis-Chalmers, *Annual Report*, 1964, p. 19.
56. *I&T*, October 1, 1962, p. 36.
57. J.I. Case, *Annual Report*, 1964, p. 3.
58. C.N.E.E.M.A., *Bulletin d'Information*, March-April, 1967, p. 61.
59. United Nations, *European Tractor Industry*, pp. 32–34.
60. This, whether justified or not, is the official view of the German industry. See Friedhelm Meier, "Die Entwicklung der deutschen Landmaschinen und Ackerschlepper-industrie," *Landmaschinen Markt*, No. 23–24, November, 1966, pp. 1483–1484.
61. United Nations, *European Tractor Industry*, p. 34.
62. *Landmaschinen Markt*, October, 1969, p. 1076.
63. *I&T*, July 7, 1951, p. 209.
64. Neufeld, *Global Corporation*, pp. 87–91.
65. *I&T*, October 6, 1956, p. 88.
66. Donaldson, et. al., *European Equipment Industry*, p. 16.
67. *L'Argus de l'Automobile*, May 28, 1964, p. 32.
68. Karl Jetter, "Die deutsche Schlepperindustrie und ihre Wettbewerber," *Landmaschinen Markt*, September 29, 1965, p. 1171.
69. "Viele von ihnen, die Renault übernehemen konnte, gingen—wie es in der Branche heisst—'mit eingerollten Fahnene und einem Trauerflor' ins französische Lager . . ." *Ibid.*
70. *Landmaschinen Markt*, November 26, 1964, p. 1526.
71. *Schlepper-Dienst*, June 1967, p. 1.
72. *Farm Engineering Industry*, September, 1969, p. 329.
73. This information was conveyed to the writer by Dr. Kiel of the Landmaschinen und Ackerschlepper Vereinigung (February 10, 1971). Another

source close to the industry reported to the writer that Deere-Lanz, rather than giving a set discount during the 1970 selling season, was prepared to print retail price lists for its dealerships with any implicit discount requested.

74. *Schlepper-Dienst*, April, 1968, p. 3; May, 1968, p. 3.

75. Barber, *Prices*, pp. 113-136.

76. *FIMR*, May 1, 1964, p. 644.

77. *Frankfurter Allgemeine Zeitung*, February 12, 1970. Similar plans were developed in 1972 in cooperation with the Saskatchewan provincial government for the assembly and sale of Universals throughout Canada and perhaps in the U.S. as well. *I&T*, February 7, 1972, p. 55. As of this writing, the project was still in the negotiating phase.

78. United Nations, *European Tractor Industry*, pp. 74-75.

79. *L'Informatore Agrario*, various issues.

80. Utenti Motori Agricole, *La Meccanizzazione Agricola in Italia* (Rome: Utenti Motori Agricole, 1951), pp. 171-183.

81. These figures are in marked contrast with those given in Joe S. Bain, *International Differences in Industrial Structure* (New Haven: Yale University Press, 1966), p. 100. In a table for which no date is given (although most of Bain's data is for the early or mid-fifties), it is implied that government controlled firms provided 67 percent of the combined output of tractors and farm machinery, and the text suggests that government controlled firms were the largest single producer in both categories. This cannot have been true for the period about which Bain was writing, and, by the sixties, the importance of the three I.R.I. firms had dwindled to insignificance.

82. Neufeld, *Global Corporation*, pp. 351-356.

83. Federal Trade Commission, *Agricultural Implement Industry*, pp. 982-983.

84. Linneman, "U.S. Tractor Industry," pp. 117-118. Chamberlain was the largest secondary industry in Western Australia at the time. *Producer's Review*, July, 1959, p. 13.

85. Linneman, "U.S. Tractor Industry," p. 199.

86. Calculated from material cited for Figure 2-1 and sales figures given in Neufeld, *Global Corporation*, p. 152.

87. *Producer's Review*, September 15, 1956, p. 30. It claimed 50 percent of all sales in the 30 h.p. and over class which was the only part of the market it was selling in; this class contained about 50 percent of 1956 sales. This can be compared with the 32 percent held by Ferguson machines in the previous year.

88. Linneman, "U.S. Tractor Industry," p. 124.

89. Chamberlain's sales for the previous years tend to confirm its rather modest position. The company claimed only 393 units sold in 1955 to 1,300 in 1959 or from less than 2 percent to 7.5 percent of total domestic sales. *Producer's Review*, September, 1959, pp. 93.

90. D.T. Brash, *American Investment in Australian Industry* (Cambridge: Harvard University Press, 1966), p. 302; *Farm Engineering Industry*, September, 1969, p. 353; *I&T*, September 9, 1969, p. 13.

91. Production estimates up to 1957 from Linneman, "U.S. Tractor Industry," p. 105. For 1957 through 1970, production was estimated as the difference be-

tween imports and total receipts (not sales, which are also recorded) to the Australian market. Detailed trade figures were made available to the author by the Acting Commonwealth Statistician, J.P. O'Neil. The receipt figures are from *Rural Industries* (Canberra: Central Office of the Commonwealth Bureau of Census and Statistics), various annual editions. The figures are considerably at variance with those given in *Manufacturing Industry* (Canberra: Central Office of the Commonwealth Bureau of Census and Statistics), various annual editions. The *Manufacturing Industry* figures apparently count the assembly of at least some of the "knocked-down" units (properly) counted in the import statistics.

92. Linneman, "U.S. Tractor Industry," p. 124.

93. Barber, *Prices,* pp. 128-136.

94. *Producer's Review,* July, 1968, p. 14.

95. *Producer's Review,* November 15, 1952, p. 53; June, 1957, p. 69; May 15, 1952, inside front cover.

96. Australian statistics divide receipts into three categories: "U.K.," "U.S." and "Other" (including Australian-made machines). Fiat's share is estimated on the reasonable assumption that it, along with Australian domestic production, virtually exhausted the "Other" category throughout.

97. Barber, *Prices,* pp. 128-136.

98. Linneman, "U.S. Tractor Industry," pp. 121, 126-127.

99. Caves, "International Corporations," p. 7.

100. Baldwin, *Non-tariff Distortions of International Trade,* p. 31.

101. International Harvester's Australian production during the fifties was quite limited and may have amounted to little more than assembly for many models. Subsequent modest production extension was a function of the subsidy system and was most unlikely to have taken place otherwise.

102. Caves, "International Corporations," p. 12.

103. Neufeld, *Global Corporation,* pp. 15-37.

104. International Harvester, *Annual Report,* various years.

105. Caves, "International Corporations," p. 4.

106. P.E.P., *Agricultural Machinery,* pp. 96-97. The figures given are for the entire British industry, but it was dominated by Ford.

107. For further discussion of production volume of these firms, see Chapters Eight and Eleven.

108. Caves, "International Corporations," p. 13. Caves does not examine sales subsidiaries because "the capital invested in them is small." There is a substantial difference in the tractor (or more broadly, farm machinery) industry, however, between a sales subsidiary, which simply gets the equipment off the dock or across the border and turns it over to others—one which, except for possible transfer price manipulation for purposes of tax avoidance may differ little from the use of an 'exclusive importer'—and the sometimes very substantial capital and current expenditures involved in the operation of a complete tractor (or farm machinery) distribution system. In the latter case, Caves's point seems quite applicable to a firm which only distributes in a foreign environment.

109. Neufeld, *Global Corporation,* pp. 88-91. It might appear that Porsche's attack on the U.S. market was a direct response to the purchase of Lanz by Deere. The move coincided with a similar attempt to become established in

Britain, however, and almost certainly resulted from an exaggerated view of the appeal of the firm's product rather than from any breakdown of spheres of influence.

110. The term is used by Bain for a similar situation in a closed economy. See *Industrial Organization*, p. 268.

NOTES TO CHAPTER EIGHT

1. Federal Trade Commission, *Agricultural Implement Industry*, pp. 268–88; Federal Trade Commission, *Manufacture and Distribution*, pp. 136–137.

2. Temporary National Economic Committee, *Investigation of Concentration of Economic Power, Monograph No. 36,* (prepared by the Federal Trade Commission) (Washington: U.S. Government Printing Office, 1940), p. 244.

3. Data in this paragraph are from *Forbes,* June 15, 1966, pp. 30–36.

4. *Ibid.,* p. 33.

5. Phillips, "The Farm Machinery Industry," p. 350.

6. *Ibid.*

7. Martinusen and Barry, *Revenues, Costs, and Profits,* p. 131.

8. In the mid-sixties the firm began to diversify into construction equipment and was quite successful in the less powerful part of the crawler market. See *Forbes,* October 1, 1968, pp. 19–20.

9. Martinusen and Barry, *Revenues, Costs, and Profits,* p. 146.

10. Due to the nature of the business, the amount of equity involved in such an operation is very low, and for this reason as well, including the amount of funds tied up in retail finance might be misleading because debt financing would be unusually easy. It was mainly to facilitate borrowing and to lower the debt ratio on main company books that the credit operations, either retail-wholesale or retail only, were sequestered off as separate subsidiaries by the major companies. See *Fortune,* February, 1958, p. 191.

11. *I&T,* January 24, 1964, pp. 30–31.

12. Neufeld, *Global Corporation,* pp. 251–254.

13. *I&T,* January 7, 1963, p. 26.

14. A.D.H. Kaplan, Joel B. Dirlam, and Robert F. Lanzillotti, *Pricing in Big Business* (Washington: The Brookings Institution, 1958), pp. 6, 69–79, 135–142; Federal Trade Commission, *Agricultural Implement Industry,* pp. 225–230.

15. Barber, *Report,* pp. 150–151.

16. *Ibid.,* p. 131.

17. Schwartzman, *Oligopoly,* pp. 121–125.

18. Barber, *Report,* p. 116.

19. Schwartzman, *Oligopoly,* pp. 108–110; Barber, *Report,* pp. 125–129.

20. Bain, *Industrial Organization,* pp. 269–276.

21. This includes a couple of truly disastrous years, 1960 and 1961, after the leadership of Marc J. Rotjman, who believed that easy credit all-around could solve Case's problems. Profit material for all firms from Moody's *Industrial Manual,* various annual editions.

22. Martin Shubik, *Strategy and Market Structure* (New York: John Wiley & Sons, 1960), pp. 304–305.

23. Kaplan, et al., *Pricing in Big Business*, p. 142.

24. Schwartzman, *Oligopoly*, p. 157.

25. The Renault position is expressed in *Farm Engineering Industry*, November, 1969, p. 410; Fiat's view was suggested to the writer by a source close to the industry.

26. Pitfield, MacKay, Ross and Company, Limited, *The Farm Machinery Industry*, Ottawa, 1969, p. 38. (Mimeographed.) "Fiat L'Offensive," *Entreprise*, December, 1970, pp. 11–17.

27. *I&T*, July 25, 1959, p. 80; February 20, 1960, p. 67; September 7, 1967, p. 36; *Farm Engineering Industry*, November, 1969, p. 410.

28. Barber, *Prices*, p. 79.

29. Neufeld, *Global Corporation*, p. 66.

30. The Justice Department and the Federal Trade Commission looked with favor upon the Hudson-Nash and Studebaker-Packard mergers with this justification at approximately the same time. Robert F. Lanzillotti, "The Automobile Industry," in Walter Adams, ed., *The Structure of American Industry* (4th ed.; New York: The Macmillan Company, 1971), pp. 263, 289.

31. Merger feelers from Deere toward the new firm in 1956 never got to a serious stage, but because Massey-Harris-Ferguson had at least 5 percent of the U.S. farm machinery market and Deere perhaps 25 percent, it is unlikely that a union between them would have been permitted by the authorities, in spite of the smaller firm's financial difficulties at the time. The reasons for Deere's interest are unrecorded. (Neufeld, *Global Corporation*, p. 200). A proposed merger between White Consolidated Industries, which held one-third of Allis-Chalmers' stock, and White Motor was blocked by the Justice Department in 1971, not because of the farm machinery link, but rather because of possible reciprocal dealing in other product lines (*New York Times*, February 26, 1971).

32. *Farm and Power Equipment*, February, 1972, p. 42.

33. Neufeld, *Global Corporation*, pp. 299–301.

34. Donaldson, et al., *European Implement Industry*, p. 13.

35. *FIMR*, May 1, 1955, p. 153; January 1, 1956, p. 1660; June 1, 1956, p. 258; August 1, 1958, p. 552.

36. This information was provided to the writer by Leslie Southcombe, Secretary of the Agricultural Machinery and Tractor Dealers' Association, Limited, in a letter dated December 16, 1970.

37. *The Economist*, June 5, 1965, pp. 1187–1188. This article is approximately correct in stressing the vital export role for the British producers, but its exact market share and export estimates are contradicted by other, more convincing, sources.

38. *FIMR*, January 1, 1965, p. 63; April 1, 1965, p. 481; January 1, 1966, p. 60; January 1, 1967, p. 44; February 1, 1967, p. 135; *AMJ*, January, 1970, p. 45; May, 1970, p. 61; September, 1970, p. 19; *Business Week*, July 17, 1971, p. 50.

39. *Farm Mechanization*, May, 1960, p. 192; Barber, *Prices*, pp. 128–136. An attempt to measure the responsiveness of market share to "quality corrected" prices (in the sense of Griliches) has been made by Cowling and Rayner for the post-war U.K. market. Although some of the assumptions made by the writers

seem rather arbitrary, they conclude that the short-run demand faced by an individual seller is (unsurprisingly) quite elastic and increased over the period. Their 1965 estimates are –5.58 in the short-run and –12.65 in the long-run. Keith Cowling and R.J. Rayner, "Price, Quality, and Market Share," *Journal of Political Economy*, 78 (December, 1970), pp. 1292–1309.

40. *AMJ*, May, 1970, p. 43; Donaldson, et al., *European Equipment Industry*, p. 13.

41. Donaldson, et al., *European Equipment Industry*, p. 13.

42. *Business Abroad*, September, 1966, p. 38.

43. *AMJ*, May, 1970, p. 40.

44. *FIMR*, April 1, 1968, p. 290.

45. C.N.E.E.M.A., *Bulletin d'Information*, March-April, 1970, p. 60.

46. *FIMR*, January 1, 1958, p. 1378; October 1, 1966, p. 1139.

47. Harvard Business School, *Deere and Company*, p. 3.

48. *I&T*, September 21, 1963, p. 35; Donaldson, et al., *European Equipment Industry*, p. 14.

49. Donaldson, et al., *European Equipment Industry*, pp. 13–14.

50. *Business Abroad*, September 5, 1966, p. 11.

51. *I&T*, September 21, 1963, p. 35; *FIMR*, July 1, 1967, p. 634.

52. *Schlepper-Dienst*, May, 1963, p. 3.

53. *I&T*, July 1, 1955, p. 38; *Farm Engineering Industry*, July, 1970, p. 218; *FIMR*, March 1, 1968, p. 204.

54. One who has in Europe's leading journalist on the subject, Karl Jetter. "Die Landmaschinenindustrie konzentriert." *Frankfurter Allgemeine Zeitung*, May 21, 1970, p. 14.

55. *Landmaschinen Markt*, October, 1969, p. 1076.

56. Donaldson, et al., *European Equipment Industry*, p. 8.

57. "Der Deutsche Traktorenmarkt gehört jetzt den Groszen," *Frankfurter Allgemeine Zeitung*, February 2, 1970. The production estimates in this paragraph are based on internal market shares and generous estimates of export sales.

58. Martinusen and Barry, *Revenues, Costs, and Profits*, pp. 179–181; *I&T* January 21, 1971, p. 43; *Wall Street Journal*, May 20, 1971, p. 15; *Business Week*, May 22, 1971, p. 28.

59. *FIMR*, April 1, 1964, p. 519; Neufeld, *Global Corporation*, p. 354.

60. *FIMR*, June 1, 1967, p. 556–57; Barber, *Prices*, p. 12.

61. *AMJ*, May, 1969, p. 9; Barber, *Prices*, pp. 128-136.

62. U.M.A., *La Meccanizzazione Agricola in Italia*, 1956, 1964.

63. Unione Nazionale Costruttori Macchine Agricole, *Trattrici* (Rome: Unione Nazionale Costruttori Macchine Agricole, 1969); Ministero del Commercio con l'Estero, *Machines Agricoles Productrices d'Energie*, (Rome: Instituto Nazionale per il Commercio Estero, 1970). Most tractor engines were produced by Slanzi, a firm which at one time produced this type of tractor.

64. Only about 100 of Lamborghini's famous luxury cars were made in 1967, so there would have been virtually no additional production economies from automobile manufacture. (*Successo*, October, 1968, p. 56). Production estimates in this paragraph are based on internal market shares and generous estimates of export sales.

65. Neufeld, *Global Corporation*, 364–373; *Producer's Review*, February, 1962, p. 4; *Power Farming Technical Annual*, 1970–71, pp. 6, 7.

66. Bernasek and Kubinski, "Agricultural Machinery," p. 476.

67. Neufeld, *Global Corporation*, p. 371; *I&T*, September 7, 1969, p. 14.

68. J.P. Nieuwenhuysen, "The Trade Practices Act: Recent Developments and Some Proposals for Change," *Australian Economic Review*, Fourth Quarter, 1969, pp. 19–24.

69. Bernasek and Kubinski, "Agricultural Machinery," p. 476; *I&T*, September 7, 1969, p. 14.

70. *Power Farm Technical Annual*, 1968–69.

71. *Producer's Review*, May 15, 1955, p. 14; October, 1965, p. 17; November, 1966, p. 83; Linneman, "U.S. Tractor Industry," pp. 112, 121; *Producer's Review*, February, 1965, p. 39.

72. *I&T*, October 21, 1970, p. 47; *Producer's Review*, November, 1970, back cover.

73. FTC, *Agricultural Machinery Industry*, pp. 1023–1026.

74. For a more detailed account of the Co-op's organization and its post-war ambitions of becoming a real competitor to the full-line companies, see FTC, *Manufacture and Distribution*, pp. 104-106.

75. Phillips, "The Farm Machinery Industry," p. 348.

76. The following discussion of CCIL is drawn from Barber, *Report*, pp. 184–188 and Simkin, *Prairie Co-operative*, especially, pp. 41–60.

77. $8 million out of $26 million of total sales in 1966 was of used machinery, and the accounting losses involved in the used equipment sales were $4 million (Can.). Simkin, *Prairie Co-operative*, p. 19.

78. Barber, *Report*, p. 188.

79. *Good Farming*, October, 1969, p. 12.

80. *AMJ*, May, 1970, pp. 40–43.

81. *FIMR*, July 1, 1961, p. 917.

82. E. Calcatierra, G. Mazzocchi, S. Lombardini, and F. Vito, "The Main Outlines of the Structure of the Italian Economy," in R. Frei, ed., *Economic Systems of the West*, Volume II (Tübingen: J.C.B. Mohr–Paul Siebeck, 1957), p. 104.

83. These figures were calculated from total sales in each market and dealer estimates from the following sources. U.S. and Canada: *I&T*, June 21, 1970, pp. 22–23; Australia: *Farm Engineering Industry*, September, 1969, p. 333; Britain and the Continent: Donaldson, et al., *European Equipment Industry*, p. 13; *FIMR*, October 1, 1966, p. 1139; *AMJ*, May, 1970, p. 43.

84. Massey-Ferguson, Limited, *Annual Report*, 1970, p. 29.

85. Agricultural Machinery and Tractor Dealers' Association, *National Survey of Trading Costs, Margins, and Profits* (Penn Place, Rickmansworth, Herts.: Agricultural Machinery and Tractor Dealers' Association, 1966); National Farm and Power Equipment Dealers Associations, *Cost of Doing Business Survey* (St. Louis, Mo: National Farm and Power Equipment Dealers Association, 1968).

86. Combine sales from *AMJ*, April, 1970, p. 9; *I&T*, April 7, 1970, p. 44.

87. *I&T*, September 21, 1967, pp. 23–31.

88. Caves, "International Corporations," p. 15.

89. Bain, *Industrial Organization*, pp. 249–250.

90. Caves, *Air Transport,* pp. 49–50.

91. Ford is treated as a single entity, and the calculation combines the output of the firms which were to become Massey-Ferguson; the latter consideration is minor because Massey-Harris's output was very small.

NOTES TO CHAPTER NINE

1. Worthington, *50 Years,* pp. 11–13, 22; *Forbes,* June 15, 1966, p. 34.

2. *FIMR,* December 1, 1951, p. 1247–1248; *I&T,* March 10, 1956, pp. 50–51; *FIMR,* December 1, 1968, pp. 1222–1223.

3. *I&T,* April 30, 1960, p. 22.

4. *FIMR,* November, 1, 1954, p. 1203; April 1, 1958, p. 1841.

5. *FIMR,* December 1, 1952, p. 1435.

6. *FIMR,* February 1, 1957, p. 1565; June 1, 1962, p. 795.

7. *FIMR,* June 1, 1967, p. 557.

8. Neufeld, *Global Corporation,* pp. 190–201.

9. *Producer's Review,* July, 1959, p. 11.

10. L.H. Hodges, "The Voluntary Standards Program for Agricultural Tractors," U.S. Department of Transportation, *Agricultural Tractor Safety on Public Roads and Farms* (hereinafter referred to as *Tractor Safety*) (Washington: U.S. Government Printing Office, 1971), pp. A125–A138.

11. Barber, *Report,* pp. 533–537.

12. Neufeld, *Global Corporation,* p. 272.

13. Russell H. Hahn, "Voluntary Standardization and ASAE," *Agricultural Engineering,* April, 1970, p. 231. Similar sentiments are expressed by engineers with great frequency: E.W. Tanquary, "Standardization: World-Wide," *Agricultural Engineering,* September, 1963, p. 486; Arnold B. Skromme, "The Growth of ASAE and the Farm Equipment Industry," *Agricultural Engineering,* April, 1970, p. 181.

14. Caves, *Air Transport,* p. 48.

15. Barber, *Report,* p. 149; White, *Automobile Industry,* pp. 213, 215.

16. *I&T,* September 21, 1964, pp. 68–69.

17. Donaldson, et al., *European Equipment Industry,* p. 12; *L'Informatore Agrario,* March, 1965; *FIMR,* March 1, 1966, pp. 320–321.

18. *Forbes,* June 15, 1966, pp. 34–35; *Farm Mechanization,* June, 1960, pp. 199–200.

19. *I&T,* August 23, 1958, pp. 70–72; Schwartzman, *Oligopoly,* pp. 238–239.

20. White, *Automobile Industry,* pp. 19–53. Thoroughgoing engineering changes, of course, can take many years. Deere's 1960 "New Generation of Power" was planned for six years and cost an estimated $40 million. Massey-Ferguson's 1965 offerings were planned over four years at a cost of $7.5 million, and Ford's 1965 line was preceded by planning of three years and apparently much greater expenditure than that of Massey-Ferguson. *Forbes,* June 15, 1966, p. 34; Neufeld, *Global Corporation,* p. 278; Phillip A. Martel, *The 1965 Ford Tractor Engine Family* (New York: Society of Automotive Engineers, Inc., 1965).

21. White, *Automobile Industry,* pp. 171–176.

22. Schwartzman, *Oligopoly,* pp. 145–147; 235–240.

23. *Landmaschinen Markt,* September, 1957, pp. 714–715.

24. *Landmaschinen Markt,* October 1969, pp. 1084–1085.

25. Martinusen and Barry, *Revenues, Costs, and Profits,* p. 122. Cf. Bain, *Industrial Organization,* p. 356.

26. Martinusen and Barry, *Revenues, Costs, and Profits,* p. 123; material on 1970 expenditures from the October, 1970 issue of *Agri-Marketing,* cited in *I&T,* December 7, 1970, p. 36.

27. Martinusen and Barry, *Revenues, Costs, and Profits,* p. 123.

28. Massey-Ferguson for a time in the late fifties sponsored a country music program on national television. "Some will argue that Red Foley and not new management put Massey-Ferguson back on its feet in America," Neufeld, *Global Corporation,* p. 255. The Foley program was, however, accompanied by selective price-cutting and other promotional efforts, so its precise impact is impossible to assess.

29. Barber, *Report,* p. 148.

30. Neufeld, *Global Corporation,* p. 13. Massey-Ferguson's Canadian market share in 1965 was 23 percent by comparison with 13.5 percent in the U.S.

31. Barber, *Report,* pp. 129–153.

32. Donaldson, et al., *European Equipment Industry,* p. 13.

33. Neufeld, *Global Corporation,* p. 264.

34. *Forbes,* November 15, 1968, p. 30–32.

35. One piece of evidence for the U.S., based on a study of 1,500 corn belt farms, shows a significant difference in costs of tractor repairs per hour used in the late sixties. Although for nine makes the range was quite broad, 18.6¢ to 55.1¢, six of the makes were in the 18.6–23.9¢ range, thus differing in average repair costs by probably less than $25 a year. The three other brands were found to average 27.2, 29.6, and 55.1¢. The brands were not identified, and because the study apparently took place over a rather short period of time, really major problems with particular new models could have influenced the results of some companies. *I&T,* May 7, 1971, p. 21.

36. *FIMR,* March 1, 1966, p. 320–321.

37. In addition, Deere made the warranty directly from the company to the customer without tying it to any particular dealer, and it also increased reimbursement to dealers for work performed. *I&T,* October 21, 1968, pp. 30–33.

NOTES TO CHAPTER TEN

1. Barber, *Prices,* pp. 17–19, 101.

2. In the U.K. prices have been reported at various times in *Farm Mechanization, AMJ,* and *Power Farming.* In France, *Motorisation Agricole, L'Argus de l'Automobile,* and *Tracteurs and Machinisme Agricole* have provided commercial publication of retail prices, and they have also been published intermittently since the mid-fifties by the government-supported C.N.E.E.M.A. in various of its publications. Australian price information for most of the period is limited to the modest price advertising done by manufacturers in *Producer's*

Review, although in the late sixties *Power Farming in Australia and New Zealand* began publishing prices in its *Technical Annual.* Italian prices for a limited and shifting set of brands have appeared regularly in *L'Informatore Agrario.* None of these sources provide completely adequate machine specification for accurate continuous international comparisons.

3. A careful discussion of discounts is presented in Barber, *Prices,* pp. 99–101.

4. Barber, *Report,* p. 159.

5. *I&T,* October 10, 1953, p. 132; Linneman, "U.S. Tractor Industry," pp. 44, 48.

6. For a small tractor in 1960, the Spearman rank correlation coefficient between height of tariff and price was .03; for a large unit –.51. In 1965, before the completion of the EEC among the original six, the figures were .11 and .03.

7. This practice is discussed in Neil B. MacDonald, "Farm Machinery Costs" (paper presented before the Edmonton Chamber of Commerce, November 13, 1970), pp. 15–16. (Mimeographed.)

8. D. Gale Johnson, *Farm Commodity Programs: An Opportunity for Change* (Washington: American Enterprise Institute for Public Policy Research, 1973), pp. 7–12, 51–71.

9. Barber, *Prices,* pp. 111–112.

10. *Ibid.,* p. 19.

11. Barber, *Report,* pp. 571–573.

12. Bain, *Barriers,* p. 314.

13. Barber, *Prices,* p. 16.

14. *I&T,* November 7, 1970, p. 13.

15. By 1970, only very rarely did the transaction completely by-pass the dealer and go straight to the manufacturer. Even when it did, however, the added savings to the buyer may have been modest. See *Ibid.*

16. N.F.P.E.D.A. *Official Guide,* Spring issues for 1961, 1966, 1970; *Farm Mechanization,* September, 1961, p. 344; *L'Informatore Agrario,* March 1966, p. 441; *Tracteurs et Machinisme Agricole,* Spring, 1970, p. 273.

17. A letter dated January 19, 1972, to the writer from the statistician of the Agricultural Engineers Association makes it clear that, while the industry was not subject to inquiry by the Prices and Incomes Board, it did engage in "early warning" of impending price rises and may well have attempted some restraint of prices.

18. These contracts apparently came into force soon after the sale of similar machines in the U.S. and U.K. became common and was used by all British firms. In 1970, after the RCFM's *Special Report on Prices,* Ford dropped the clause from its English dealer contracts but still had an important means of surveillance; too many tractors taken by dealers in a virtually stagnant market would become quickly apparent to the manufacturer. Barber, *Report,* p. 159.

19. *I&T,* June 7, 1970, pp. 12–15; *Farm and Power Equipment,* December, 1970, pp. 32–34.

20. Deere, some of whose German machines made their way into the U.S. market through bootleg channels after 1960, moved in 1969 to make its warranty directly on the machine rather than through a specific dealer. It apparent-

ly valued its reputation for standing behind its product more than any possible deterrent on imports caused by the no-warranty situation. *Canadian Farm Implement Dealer,* August, 1969, p. 18.

21. *I&T,* June 7, 1970, p. 13.

22. The structure of international prices for farm machinery was a subject of the massive FTC study of 1938, but in the pre-war period prices were higher abroad almost without exception, at least for the products of American firms. For the single post-war exception to the sustained U.S. inattention to international price comparisons, see p. 219.

23. See Canada, House of Commons, Standing Committee on Agriculture and Colonization, *Minutes of Proceedings and Evidence,* Nos. 1–13, 24th Parliament, 3rd Session, 1960.

24. See the Order in Council (Privy Council) printed at the beginning of Barber, *Prices.*

25. *Good Farming,* August, 1969, pp. 18–31.

26. *Canadian Farm Equipment Dealer,* January, 1969, pp. 16–17.

27. *Ibid.,* p. 7; *I&T,* February 21, 1970, p. 12.

28. Barber, *Prices,* p. 91.

29. *Canadian Farm Equipment Dealer,* March, 1970, p. 7.

30. *Canadian Farm Equipment Dealer,* August, 1969, p. 18, reported that "a few of the more radical groups" on the U.S. West Coast were importing tractors directly from abroad, but the writer could find no verification for this assertion. A letter dated January 24, 1972, to the writer from the program director of the Ontario Federation of Agriculture (OFA) said that he was unaware of any farm group in the U.S. pursuing lines of action similar to those of the OFA.

31. *FIMR,* March 1, 1961, p. 353; September 1, 1963, p. 1238.

32. *FIMR,* July 1, 1965, p. 855; *AMJ,* May, 1970, p. 11.

33. In any event Kennedy Round tariff cuts moved the nominal rate from 15 percent to 5.5 percent on January 1, 1972, although Britain's entry into the enlarged Common Market will mean an ultimate tariff level to non-members of 18 percent.

34. *FIMR,* February 1, 1962, p. 216; *I&T,* May 7, 1963, pp. 26, 27, 60, 61.

35. *FIMR,* April 1, 1954, p. 2089; February 1, 1955, p. 1790; April 1, 1960, p. 1729; November 1, 1965, p. 1406.

36. *Producer's Review,* May 15, 1955, p. 14.

37. *Producer's Review,* October, 1965, p. 17; November, 1966, p. 83.

38. Heady and Tweeten, *Resource Demand,* pp. 91–93; Scott and Smyth in *Demand for Farm Machinery,* pp. 71–72, make a similar argument.

NOTES TO CHAPTER ELEVEN

1. Neufeld, *Global Corporation,* pp. 249–289; pp. 303–329.

2. *Ibid.,* pp. 190–201.

3. Massey-Ferguson Limited, "Statement on the Special Report on Prices of the Royal Commission on Farm Machinery," Toronto, March 3, 1970, p. 4. (Mimeographed.)

4. Neufeld, *Global Corporation,* p. 256.

5. Barber, *Report,* p. 154.

6. This is based on Ford's estimated U.S. share and Ford's exports to the U.S. reported in *I&T,* September 3, 1960, p. 76.

7. *Forbes,* December 1, 1960, p. 32.

8. *Financial Post,* May 15, 1965; Schwartzman, *Oligopoly,* p. 80.

9. MacDonald, *Locational Advantages,* p. 68; *I&T,* March 7, 1965, p. 22.

10. Barber, *Prices,* p. 37.

11. *FIMR,* November 1, 1964, pp. 1485-1486; *Forbes,* November 15, 1968. pp. 30-32.

12. International Harvester Company, *Annual Reports,* various years.

13. International Harvester Company, *Annual Report,* 1964, pp. 2-3; *FIMR,* July 1, 1967, p. 653.

14. *Producer's Review,* May 1960, p. 83.

15. *I&T,* May 16, 1959, pp. 90-91.

16. *Forbes,* September 15, 1962, pp. 21-22.

17. MacDonald, *Locational Advantages,* p. 73.

18. *Forbes,* June 15, 1966, p. 31.

19. *Business Week,* July 11, 1970, p. 80.

20. Remarks of William A. Hewitt, President of Deere and Company, before the New York Society of Security Analysts, February 14, 1957. (Privately printed by Deere and Company.); *Forbes,* March 1, 1961, pp. 22-23; remarks by Clifford L. Peterson, Vice-President of Deere and Company, before the American Industrial Development Council Conference, St. Louis Missouri, April 5, 1965. (Mimeographed.)

21. Peterson, remarks cited in previous footnote.

22. Deere and Company, *Annual Report,* 1964, p. 7.

23. Barber, *Prices,* p. 9.

24. *Business Week,* October 27, 1973, pp. 77-78.

25. *Power Farming,* May, 1971, p. 63.

26. *Business Week,* February 12, 1972, p. 32.

27. *Ibid.*

28. *Tracteurs et Machines Agricoles,* Spring, 1970, p. 274.

29. Donaldson, et al., *European Equipment Industry,* p. 8.

30. This information is based on a conversation with a former management employee in Renault's agricultural machinery operations.

31. *FIMR,* March 1, 1966, p. 321.

32. *I&T,* August 7, 1972, p. 47; October 21, 1973, p. 17.

33. *FIMR,* November 1, 1967, p. 1037; *AMJ,* January, 1970, p. 45.

34. White Motor Corporation, *Annual Report,* 1966, 1969.

35. Barber, *Report,* p. 160, notes the negotiations but identifies the firm only as "a large North American company."

36. *I&T,* March 21, 1972, pp. 20-21.

37. *FIMR,* September 1, 1965, p. 1111; May 1, 1967, p. 453; *I&T,* September 7, 1963, p. 54; May 7, 1971, p. 1.

38. Barber, *Report,* p. 160.

39. Caves, "Foreign Investment," p. 15.

40. *Ibid.,* p. 46.

41. Deere's American-designed tractors for Germany and diesel engines for France have already been noted. For the Massey-Ferguson experience, see Neufeld, *Global Corporation*, pp. 276–281. Despite the fact that Ford's 1965 tractors were to be overwhelmingly produced at Basildon, all original research and initial design of their engines (the area of most thoroughgoing revision) was done in the U.S. This was presumably due to the presence of greater engineering capability in Detroit than abroad, a capability which will presumably be maintained to concentrate on increasingly large models. See Martel, *The 1965 Ford Engine Family*.

42. Raymond Vernon, "International Investment and International Trade in the Product Cycle," *Quarterly Journal of Economics*, 80 (May, 1966), pp. 190–206.

43. See FTC, *Agricultural Machinery Industry*, pp. 52–58.

44. This surmise is supported by the destination of the machinery and the discussion in Barber, *Report*, pp. 273–274.

45. Canadian export activity, reflected in Figure 11-1, was almost entirely directed to the U.S. and represented almost solely the activity of Cockshutt between 1947 and the firm's purchase by White in 1962.

46. Bernasek and Kubinski, "Agricultural Machinery," p. 492.

47. All following discussion of export destinations is based on the trade material cited for Figure 11-2.

48. The evidence is based largely on their market shares outside the country of manufacture, including those in countries not included in the present study. For examples of markets in the latter category, see Donaldson, et al., *European Equipment Industry*, p. 10.

49. Barber, *Prices*, pp. 133–136.

50. *Successo*, April, 1968, p. 100.

51. *Ibid.*

52. Staffan Burenstam Linder, *An Essay on Trade and Transformation* (New York and Stockholm: John Wiley & Sons, 1961), p. 105. The thrust of Linder's approach can be captured only by reading (at the very least) all of his discussion of trade in manufactures, pp. 87–109.

53. *Ibid.*, p. 94.

NOTES TO CHAPTER TWELVE

1. Barber, *Report*, pp. 128–129.

2. Martinusen and Barry, *Revenues, Costs, and Profits*, p. 10.

3. Barber, *Report*, p. 205.

4. *American Machinist*, 112 (1968), p. 35.

5. U.S. Bureau of the Census, *Annual Survey of Manufactures* (Washington: U.S. Government Printing Office), various annual issues; U.S. Department of Labor, *Handbook of Labor Statistics*, (Washington: U.S. Government Printing Office), various annual issues.

6. The case of a sky-rocketing accounting rate of return as the book value of equipment sinks toward zero with net revenue constant is only the most intuitively obvious difference between the accounting rate and the internal rate

of return. Even if investment is in a "steady state" in the sense of a firm holding and renewing a capital stock of fixed age structure, the internal rate of return and accounting measures will generally diverge. The way in which they do depends inter alia on depreciation and expensing policy, average productive life of assets, amount of working capital, gestation period of fixed investment, and the pattern of cash flow. The simplest exposition of the considerations can be found in Ezra Solomon, "Alternative Rate of Return Concepts and Their Implications for Utility Regulation," *Bell Journal of Economics and Management Science,* Spring 1970, pp. 65–81. For a more sophisticated general treatment, see Thomas R. Stauffer, "The Measurement of Corporate Rates of Return: A Generalized Formulation," *Bell Journal of Economics and Management Science,* Autumn, 1971, pp. 434–469.

7. Barber, *Report,* p. 207.

8. *Forbes,* June 15, 1966, p. 30.

9. Material on both machine sizes and manufacturing costs from *Business Week,* October 27, 1973, pp. 74–78.

10. *Forbes,* November 15, 1968, p. 30.

11. *Ibid.,* p. 31.

12. The reader is referred to Schwartzman, *Oligopoly,* pp. 77–82, for the detailed assumptions upon which these estimates are based.

13. The data upon which the transfer price calculation is made are presented in MacDonald, et al., *Tractor Costs.* The value of U.S. wholesale tractor shipments is given in U.S. Bureau of the Census, *Current Industrial Reports,* Series M35S (Washington: U.S. Government Printing Office).

14. *Il Trattorista,* April 30, 1969, p. 155.

15. C.N.E.E.M.A., *Bulletin d'Information,* March–April, 1970, p. 62.

16. *Forbes,* September 15, 1962, p. 21.

17. One is tempted to extend the concept of "X-inefficiency," defined by Leibenstein as a violation of the assumption that "every firm purchases and utilizes all of its inputs 'efficiently'," to cover cases of the failure of firms to move to lower cost sources of production. The problems with organized labor (and much of management) discussed in Chapter Eleven render this extension questionable. Nevertheless, both X-inefficiency as usually treated and the failure of firms to seek areas of low absolute cost inhere in the discretion characteristic of oligopoly with impeded entry. See Harvey Leibenstein, "Allocative Efficiency vs. 'X-Efficiency,'" *American Economic Review,* 61 (June, 1966), pp. 392–415.

18. For a discussion of the conditions which must be fulfilled for an international industry evaluation to rest firmly upon the theory of comparative advantage, see Mahlon R. Straszheim, *The International Airline Industry* (Washington: The Brookings Institution, 1969), pp. 150–154. An examination of the way in which resource savings attenuate "second best" arguments as usually formulated is presented by William S. Comanor and Harvey Leibenstein, "Allocative Efficiency, X-Efficiency, and the Measurement of Welfare Losses," *Economica,* 37 (August, 1969), pp. 304–309.

19. All of the figures are for illustrative purposes only, but it would appear that the RCFM's use of 54 percent of Suggested Retail Price (an unweighted

average of J.I. Case and Deere's cost of goods sold for 1957–1967) probably overestimates tractor costs, because tractors are generally regarded as a relatively high profit item (50 percent is 1 percent lower than Deere's average over the period.) Maule claims that 65 percent of SRP for a weighted average value of shipments figure is accurate within two percentage points. Maule, *Productivity in the Farm Machinery Industry*, p. 11.

20. This is simply the retail price minus the dealer discount of 18 percent.

21. This method of comparison is modelled on Barber, *Prices*, pp. 169–180.

22. N.R.F.E.A., *Official Guide*, 1961, p. 98.

23. Barber, *Prices*, pp. 182–200.

24. MacDonald, *Locational Advantages*, p. 87.

25. These are simply conditions unfavorable to the efficacy of the "purchasing power parity" hypothesis. See R.E. Caves and R.W. Jones, *World Trade and Payments* (Boston: Little, Brown and Company, 1973), pp. 335–338.

26. *The Federal Register*, Volume 37, No. 86, pp. 8943; *Business Week*, May 13, 1972, p. 122.

27. This ignores the small and minimally integrated facility of Versatile completed in 1966. MacDonald, *Locational Advantages*, p. 73.

28. *I&T*, September 21, 1957, p. 48; *FIMR*, August 1, 1957, p. 572; January 1, 1966, p. 60.

29. *Annual Reports* of companies.

30. International Harvester, *Annual Reports*.

31. Schwartzman, *Oligopoly*, p. 81, and International Harvester, *Annual Reports*.

32. *Forbes*, September 15, 1962, p. 25.

33. Some modernization of facilities appears to have been virtually forced upon the company in the late sixties because of inadequate facilities to produce large tractors. *I&T*, May 21, 1967, p. 69.

34. *FIMR*, June 1, 1965, p. 763; January 1, 1965, p. 63; January 1, 1967, p. 44; *I&T*, January 9, 1960, p. 71.

35. This view is confirmed in a careful outline of technical developments by Bruce Narsted, "Tractor Technical Development," in Vicas, *Research and Development*, pp. 75–98.

36. Joseph A. Schumpeter, *Capitalism, Socialism, and Democracy* (London: Unwin University Books, 1954), p. 82.

37. Deutz and Ford did this in an effective way by simplifying engine design so that the same block and many engine components could be used throughout most of the line. *FIMR*, June 1, 1966, p. 660; Martel, *The 1965 Ford Engine Family*.

38. *I&T*, November 21, 1964, p. 31.

39. *I&T*, September 21, 1957, p. 49.

40. *FIMR*, November 1, 1957, p. 1.

41. *I&T*, May 4, 1957, p. 44.

42. This issue is examined in the context of the Chamberlin large group by R.L. Bishop, "Monopolistic Competition and Welfare Economics," in *Monopolistic Competition Theory: Studies in Impact*, R.E. Kuenne, ed. (New York: John Wiley & Sons, Inc., 1967), pp. 255–260.

43. *FIMR*, March 1, 1957, p. 1.

44. *I&T*, September 21, 1970, pp. 16–20.

45. Donaldson, *Farm Machinery Safety*, p. 51.

46. *Ibid.*, pp. 50–51.

47. *I&T*, March 21, 1970, pp. 18–20.

48. Department of Transportation, *Tractor Safety*, p. 39.

49. *FIMR*, September 1, 1962, p. 1228; May 1, 1966, p. 1; *Landmaschinen Markt*, March 1970, pp. 198–200.

50. *I&T*, January 1, 1971, p. 26.

51. Department of Transportation, *Tractor Safety*, Appendix B.

52. *Ibid.*, p. 41.

53. *Ibid.*

54. *Ibid.*, pp. B5–B31.

55. *Ibid.*, p. 41.

56. Compare Chrysler's sales with estimates for world-wide farm machinery sales presented in Harvard Business School, *A Note on the Farm Equipment Industry*, ICH 13G122 BP 864R2, p. 41.

57. Barber, *Report*, p. 179.

58. *Ibid.*, pp. 166–168.

59. Martinusen and Barry, *Revenues, Costs, and Profits*, p. 103.

60. Schwartzman, *Oligopoly*, p. 26.

61. Martin A. Abrahamsen, "Discussion: The Changing Structure of Markets for Farm Machinery," *Journal of Farm Economics*, 40 (December, 1958), p. 1184.

62. The figures in Table 12-6 are weighted averages of the branch expenses experienced by all full-line companies in Canada for the only two years for which the RCFM collected data. The degree to which the difference between the two years reflects the demise of smaller dealers over time, and the consequent transaction and inventory economies, against the relative prosperity of the industry in 1966 cannot be discerned.

63. By convention, distribution advantages are usually treated as a product differentiation barrier. Nevertheless, when scarce and tangible distribution resources are involved, it would seem that they are analytically similar to scarce manufacturing inputs, control over which confers an absolute cost barrier against potential entrants. The standard treatment was developed by Bain; see *Industrial Organization*, pp. 260–261.

NOTES TO CHAPTER THIRTEEN

1. Letter to the writer from Professor Varden Fuller, Department of Agricultural Economics, University of California, Davis.

2. House Committee on Agriculture, *Farm Price-Cost Squeeze*, especially pp. 434, 435, 544, 545.

3. The relation of discriminating monopoly prices to a single price is developed in Robinson, *The Economics of Imperfect Competition*, Chapter 15.

4. For a discussion of the interpretation given the Article by the Commission and the Court of Justice during the sixties, see Grant W. Kelleher, "The

Common Market Antitrust Laws: The First Ten Years," *Antitrust Bulletin,* 12 (Winter, 1967), pp. 1219–1252. The unofficial English translation of Article 85-1 given in the text is from this source.

5. *FIMR,* March 1, 1961, pp. 333.

6. Barber, *Report,* p. 255.

7. *Ibid.*

8. In Britain large advertising outlays in the soap industry have been successfully attacked by the Monopolies Commission, and some have advocated a similar move by U.S. authorities. Scherer, *Market Structure,* pp. 344–345.

9. Barber, *Report,* p. 199.

10. *Ibid.,* p. 255.

11. Barber, *Prices,* pp. 97–98. For a critique of the proposal, see Robert T. Kudrle, "A 'Reverse Dumping Duty' for Canada?" *Canadian Journal of Economics,* 7 (February, 1974), pp. 75–81.

12. For a discussion of the "Meeting Competition Defense," see Eugene M. Singer, *Antitrust Economics* (Englewood Cliffs, New Jersey: Prentice-Hall, Inc., 1968), pp. 232–236.

13. See reference cited in footnote 3.

14. This discussion ignores the theoretical possibility that banning discrimination could cause the industry to sell only in the market with high prices under discrimination. See reference cited in footnote 3.

15. E.F. Wigglesworth, ed., *Readings in Policy and Practice in International Business* (New York: Thomas Ashwell & Company, Inc., 1961), p. 148.

16. In 1961, in the U.S. *Official Guide,* International Harvester claimed a price of $2,744 for its B-275 and noted this was "F.O.B. factory in England," The suggested retail price for English buyers was $1,556.

17. For a penetrating discussion of cooperatives written at the time of the movement's apogee of influence in the U.S., see William N. Loucks and J. Weldon Hoot, *Comparative Economic Systems* (4th ed.; New York: Harper & Bros., 1952), pp. 711–815.

18. *Producer's Review,* May 15, 1955, p. 15.

19. Brash, *American Investment in Australian Industry,* pp. 126–127.

20. Caves ("Foreign Investment," p. 42) has pointed to Australia as a country in which the protection of domestic manufacturing is in part motivated by the desire to shift the distribution of income toward labor and away from the owners of natural resources. Whatever the merits of the policy with respect to manufacturing in general, one strongly suspects that tractor manufacture would come a long way down the list of economically promising candidates for import substitution.

21. That gross production should represent a lower bound to the valuation of a life is argued strongly by A.R. Prest and Ralph Turvey, "Cost-Benefit Analysis, A Survey," *Economic Journal,* 75 (December, 1965), p. 723. The combined average on and off farm income per farm in 1970 (with government payments subtracted) was $10,936. United States Department of Agriculture, Economic Research Service, *Farm Income Situation,* July, 1971, pp. 72–73. Using this as the value of gross (i.e., gross of consumption) production does not take into account the extent to which the bulk of farm work may be per-

formed by more than one person. On the other hand, it also ignores the extent to which families deprived of the head of household would subsequently be dependent on state aid. Needless to say, the costs of pain, family anguish and medical treatment could also be extremely important.

22. Department of Transportation, *Agricultural Tractor Safety on Public Roads and Farms,* pp. 144–145.

23. Barber, *Report,* pp. 536–537.

24. *AMJ,* March, 1970, p. 9.

25. It could also conflict with maximizing the total income of the country against whose residents discrimination is being practiced if the firms practicing discrimination are owned there. See Kudrle, "A 'Reverse Dumping Duty' for Canada?," p. 80.

26. This ignores the rapidly increasing use of tractors in developing countries, but this seems a justifiable omission for at least the near future, because most such tractor use seems to be taking place *pari passu* with the creation of sub-optimal plant capacity which is given heavy protection.

27. Williamson's original formulation of the problem was overly simplified and was followed by two published corrections. See Oliver E. Williamson, "Economies as an Antitrust Defense," *American Economic Review,* 58 (March, 1968), pp. 18–36; "Correction and Reply," *Ibid.,* 58 (December, 1968), pp. 1372–1376; "Economies as an Antitrust Defense: Reply," *Ibid.,* 59 (December, 1969), pp. 954–959.

28. An elaboration of monopsony working against monopoly (or oligopsony against oligopoly or some combination) is found in J.K. Galbraith, *American Capitalism, The Concept of Countervailing Power* (Cambridge; The Riverside Press, 1952), where such interactions are illustrated by historical examples. While the possible beneficent impact of such power balance is obvious, far less acceptable is the part of Galbraith's argument which suggests that there is a natural tendency for market power assymetries between buyers and sellers to be self-correcting as organization is induced as a self-defense measure in the more fragmented side of the market. Indeed, the farm machinery industry and its farmer-customers in most countries provide an example of cases where such a development has been negligible. On the incentive for collective action, see p. 168.

Index

About the Author

Robert Thomas Kudrle is Assistant Professor of Public Affairs at the University of Minnesota and has served as Assistant Director and Acting Director of the University's Harold Scott Quigley Center of International Studies. He received his Ph.D. degree in economics from Harvard University where he was a graduate research associate at the Center for International Affairs. He also holds the B.Phil. in economics from Oxford University and the A.B. in government from Harvard College. Dr. Kudrle is the author and co-author of several articles on industrial organization, the multinational enterprise, and the political economy of social services.